# Electronics for the Beginner

by

J. A. Stanley

Howard W. Sams & Co., Inc.
4300 WEST 62ND ST. INDIANAPOLIS, INDIANA 46268 USA

Copyright © 1960, 1964, 1966, 1968,
and 1980 by Howard W. Sams & Co., Inc.,
Indianapolis, Indiana 46268

THIRD EDITION
SECOND PRINTING—1982

All rights reserved. No part of this book shall be
reproduced, stored in a retrieval system, or
transmitted by any means, electronic, mechanical,
photocopying, recording, or otherwise, without
written permission from the publisher. No patent
liability is assumed with respect to the use of
the information contained herein. While every
precaution has been taken in the preparation of
this book, the publisher assumes no responsibility
for errors or omissions. Neither is any liability
assumed for damages resulting from the use of
the information contained herein.

International Standard Book Number: 0-672-21737-6
Library of Congress Catalog Card Number: 80-51718

*Printed in the United States of America.*

# Preface

Today we live in a world of electronics: we cook in microwave ovens; we do simple arithmetic with pocket-size calculators; we watch television bounced off satellites parked high above the earth; our children play with electronic toys; and, finally, we use computers to keep records on almost everything we do in a lifetime.

Because the various facets of electronics are so much a part of modern life, there is a real advantage to having some knowledge of the art. In the writer's opinion—based on decades of experience in the electronics field—building and using simple equipment is the ideal first step to entering the world of electronics, whether it is to be a hobby or a profession.

The pleasant thing about the "build equipment" approach is that you learn without even being aware that you are acquiring such basic knowledge as a vocabulary of electronic terms, a "feel" for what happens as a signal moves through a circuit, and an understanding of servicing techniques. You can accomplish all this without spending one minute on dry theory (that can come later), and theory will be much easier to comprehend when you have seen the practical application *first*. By following through the sequence of projects in this book, you will acquire knowledge the fun way, by doing instead of by simply reading about abstract ideas.

Each project both tells and shows you everything you need to do, step-by-step, wire-by-wire, in order to build the unit described. It is assumed that you know nothing about electronics, so the first project is the simplest radio that it is possible to build. The projects become more complex as you move through the book. The final project is a shortwave receiver that, despite its simplicity, will pull in stations from all over the world. And, the book includes a chapter covering an easy way to make your own printed circuits, the technique used in so much modern electronic equipment.

There is no pretense that the projects in this book are the last word in electronic design. Rather, the circuitry chosen in each case is the *simplest* the writer could develop for a given set. Yet, each of the units is capable of good performance for its intended use, even though it is built with a minimum of parts and limited wiring. Keeping things simple helps ensure that even a beginner will build successful equipment.

Finally, because the projects are simple and easy, they are fun to build. So, whether you are a teenager who wants to try your hand at electronics or a senior citizen who is interested in embarking on a fascinating hobby, let's acquire a few tools and parts and build a radio!

J. A. STANLEY

*To Anne, whose patience and help made this book possible.*

# Contents

### CHAPTER 1

TOOLS AND THE GENTLE ART OF SOLDERING . . . . . 9

    Tools for Holding and Bending—Long-Nose Pliers—Tools for Turning—Tools for Cutting and Drilling—Tools for Soldering—The Gentle Art of Soldering

### CHAPTER 2

BUILDING A TRAP FOR RADIO WAVES . . . . . . . 19

    The Inverted L Antenna—Trees Are Quite Handy—Mounting the Mast—The Wire Itself—How to Outwit Lightning—House Breaking Made Easy!—Look Mom, No Connections!—For Apartment Dwellers Only

### CHAPTER 3

THE TWO HOUR RADIO . . . . . . . . . . . 32

    How It Works—To the Moon and Back—Let's Start Building—Poor Antenna, Weak Signals?

## CHAPTER 4

SELECTIVE AM TUNER . . . . . . . . . . 46

A Technique to Learn—Inserting Clips—Wiring—Testing—Want to Use a Loudspeaker?—Poor Antenna?

## CHAPTER 5

LOUDSPEAKER AMPLIFIER . . . . . . . . . 54

Preparing the Baseboard—Wiring—Adding Parts—Loudspeaker—A Word About Selecting Speakers—Checking the Amplifier—Troubleshooting

## CHAPTER 6

TRANSISTOR AM RECEIVER . . . . . . . . 67

Construction—Testing the Unit—No Whistles?

## CHAPTER 7

HI-FI AM TUNER . . . . . . . . . . . . 77

Wiring—Testing—Assembly

## CHAPTER 8

ABOUT PARTS AND SYMBOLS . . . . . . . . 87

Late American Sign Language—Antennas—Grounds—Wires—Carbon Resistors—Wirewound and Variable Resistors—Capacitors—One Millionth of a Millionth—Variable Capacitors—Coils—Transformers—Diodes and Transistors—FET Units—Integrated Circuits—Switches—Phones—Tubes—Other Common Parts

## CHAPTER 9

READING CIRCUITS . . . . . . . . . . . . 102
    Read From Left to Right—Applying What You Have Learned

## CHAPTER 10

UTILITY AMPLIFIER—FIRST STEP TO STEREO . . . . 108
    Built Around One IC—Wiring the Board—Auxiliary Parts—Add a Signal—Testing—Finishing Up—Why Two Phono Jacks?

## CHAPTER 11

TWO TUNERS: CB AND AIRCRAFT . . . . . . . 117
    Inserting the Terminals—Coils and Chokes—Adding Parts and Wiring—Converting to Aircraft—Into the Cabinet—Antenna and Antenna Coupling

## CHAPTER 12

SHORTWAVE RECEIVER . . . . . . . . . . 127
    Collecting the Parts—Wiring the Detector Board—Winding the Coil—Preparing the Chassis and Panel—Assembly—Testing the Amplifier—Completing the Set—Final Checkout—Learn by Practice—Some Fine Points—Tuning for CW—SSB, Also—Troubleshooting

## CHAPTER 13

MAKING YOUR OWN PRINTED CIRCUITS . . . . . . 144
    Making Your Own—The Simplest Approach—Applying the Pattern to the Board—Etching the Board—How to Know When You Are Done—Final Cleanup—Wiring a Circuit Board

## CHAPTER 14

PARTS AND PARTS SUBSTITUTES . . . . . . . . 154
    Overall Considerations—Considerations by Project—What to Do if You Cannot Find Parts Locally

**CHAPTER 1**

# Tools and the Gentle Art of Soldering

The thrill of taking a few electronic parts and putting them together to create a piece of equipment which comes alive with sound is one you will remember all your life. Like most good things, however, it has some pitfalls. One of these is the selection and use of tools needed to do the job. So the purpose of this chapter is to help you decide what tools to buy and how to use them.

Tools actually provide you with specialized "hands" for doing what would be impossible without them. For electronics work, literally hundreds of tools have been designed, many of them specialized types good for only one task. But, to get started, you will want tools which serve several purposes.

### TOOLS FOR HOLDING AND BENDING

You probably already have some type of slip-joint pliers—the usual hand pliers which can be opened to handle fairly large objects. If you are buying a new pair, select one with a fairly thin "nose," because you often will be working in crowded quarters.

Always buy *good* quality tools because they *are* your hands for much of the building you will do. Poor quality tools make any job difficult. They may even make it impossible. Further-

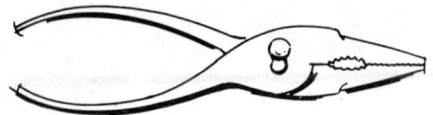

more, tools—good ones—are virtually a lifetime investment. The writer is still using (after more than thirty years) several pliers he bought while in college.

## LONG-NOSE PLIERS

The slip-joint pliers mentioned are useful for such jobs as holding nuts and bending fairly large pieces of metal. However, for a lot of electronics work they are too big. The tool needed is a pair of long-nose pliers. The six-inch size will be large enough for most of your work; they need not have cutters. Long-nose pliers are highly useful, not only for slipping into places you can't reach with ordinary pliers, but also for

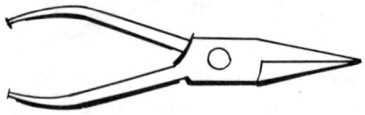

forming wire into loops. As an example, you can use them to loop the end of a light cord when you are attaching the cord to a plug or socket.

## TOOLS FOR TURNING

Of course you will need screwdrivers. Although good quality is advisable, the medium-priced ones will do the job well enough to get you started. Actually, you will need three sizes:

one "pocket" size, small enough to use on small setscrews in knobs; one medium size for ordinary machine and wood screws; and one fairly husky screwdriver. The third one won't see much use, but it will be a godsend when you tackle a "toughie," like removing a rusted bolt from some ancient radio you acquired for salvage.

The two pairs of pliers mentioned will take care of most of your needs. However, a pair of special socket wrenches—miniature versions of the cross wrenches used in removing wheel lugs to change a tire—will be extremely useful. Each tool has

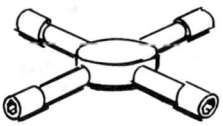

four sockets, giving a total choice of eight—just about all you will ever need for working on most electronic equipment. These wrenches do a better job of turning or holding nuts than do pliers, yet they are not nearly as expensive as a set of socket wrenches.

## TOOLS FOR CUTTING AND DRILLING

A crack tv service technician of my acquaintance says his most important service instrument is a pair of diagonal cutting pliers, which he uses to clip out bad parts. You won't be using them in this fashion for awhile, but you will use them constantly for snipping off ends of wires. They also make excellent wire strippers, and you will not need to buy an additional tool for this purpose. The trick is to use your forefinger as a spacer to prevent cutting the wire while stripping off the insulation. Study the drawing on the next page, and then try the technique on a length of insulated wire. It takes a little practice; you'll probably cut the wire the first time!

In buying diagonal cutting pliers, again buy quality. In fact, if you are buying only one quality tool, let it be the diagonal

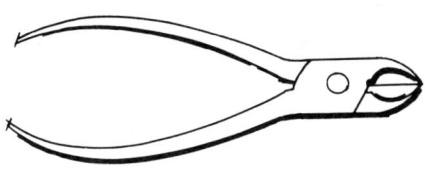

11

cutters. Since this tool is used for cutting, it will pay you to get the finest tool steel. This means the cutters may cost ten dollars. A less expensive pair will soon nick along the cutting edges.

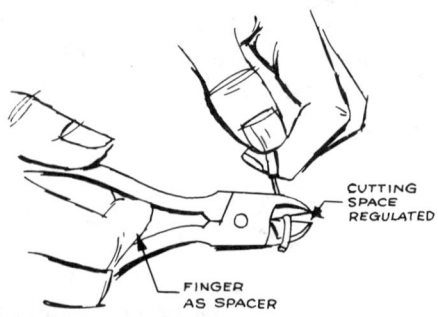

CUTTING SPACE REGULATED

FINGER AS SPACER

If your budget permits, a tool made specifically for stripping wires is highly useful. Unlike the versatile diagonal pliers, a wire stripper serves only one purpose, but it is easy and handy to use.

For sawing off shafts and cutting large holes in metal, you will need some type of hacksaw. Since the blades are usually of the same quality, regardless of the type of frame, a low-priced keyhole hacksaw will do a satisfactory job.

For drilling holes, a drill is a must. It may be either a hand drill or a power drill (if you can handle it). Buy the best one your budget allows. A set of drill bits, from $\frac{1}{16}$ to $\frac{1}{4}$ inch, is

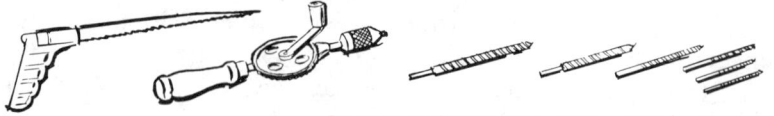

BE SURE THE BITS WILL DRILL METAL!

ideal. However, just two drill bits, $\frac{5}{32}$ and $\frac{1}{4}$ inch, will handle a surprising amount of building. For larger holes, the drill you will need most often is the $\frac{1}{2}$-inch size. You can buy one with a shank that fits into an ordinary carpenter's brace.

Some type of pocket knife is needed, too. A good Scout knife is ideal because it includes an *awl*, or *reamer*, blade that is also highly useful.

## TOOLS FOR SOLDERING

Yes, you will need a soldering iron. But note the word *iron*, not gun. The latter, although ideal for certain work, can get a beginner into real trouble. Unless used skillfully, it will become quite hot if not temperature controlled. The result can be over-

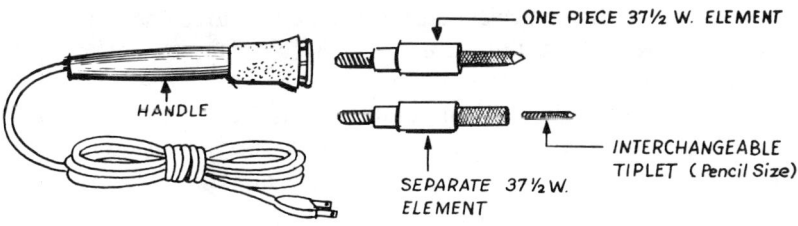

cooked solder joints or damaged small parts. For this reason, it is far better to start off with a soldering iron. Besides, the iron is less expensive.

An excellent iron to buy is one of the "pencil" types with interchangeable elements. This feature makes it possible for you to buy only one handle, and then screw in the different tips for handling light or medium work. For example, the handle can be fitted with a 37½-watt pyramid-tip element, which is adequate for fairly heavy work when the right solder is used. For light work or in tight quarters, the pyramid-tip element can be removed, and a 37½-watt element fitted with a pencil-type "tiplet" used instead. This "tiplet," about the diameter of a lollipop stick, will enable you to work in mighty small spaces.

For certain types of soldering (for example, printed circuits), a lower-wattage element may be desirable. More will be said about that in a subsequent chapter.

In addition to the iron, another tool, usually called a "soldering aid," will be a tremendous help. It is a piece of stainless-steel rod fitted with a wooden handle. One end of the rod is slotted, for gripping and twisting wire. The other end is pointed, for poking hot solder out of holes in lugs or other connectors. Because the rod is stainless steel, solder does not readily stick to it.

## THE GENTLE ART OF SOLDERING

Now, for soldering—the basic art in all electronic building. You *must* learn to solder and to solder *well*. More beginners have trouble with soldering than with any other phase of building.

Some new irons will need "tinning" before they are used. Because some irons are already tinned or require special treatment, be sure to follow the manufacturer's instructions. If none are given, do the following: File the tip of the iron until it shines. Then heat the iron, and coat the tip with solder. Tinning is the *only* time you apply solder *directly* to the soldering iron. More—a LOT more—on that subject later.

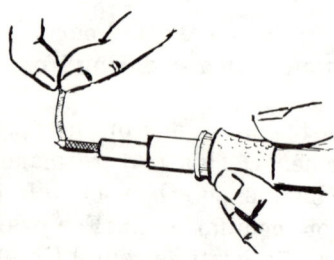

One of the best ways to learn any process is to study an actual example, so let's do exactly that. We'll start by soldering a wire to a terminal lug—something you will be doing frequently.

First, strip back the wire, using the "finger-in-the-diagonal" technique described earlier. But don't stop there. With a piece of sandpaper (or the blade of a knife), scrape the wire lightly until its surface is bright and as free as possible from oxida-

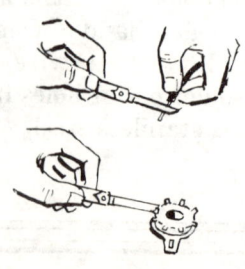

tion (which might cause a bad solder joint). Do the same with the lug, even though it looks shiny.

*Remember* this step. *Scrape* both parts or wires to be joined, making certain the solder has a clean surface to adhere to.

It is true, of course, that many parts you buy are already "tinned." Theoretically, with such parts you can skip the "scraping" step. However, until you have had considerable experience with soldering, the safest way is to scrape *everything*.

The next step is to join the parts mechanically so that all the solder has to do is make the electrical connection. Don't think of solder as a kind of glue—that isn't its purpose. Never, NEVER depend on solder alone to make the mechanical joint.

In our example of hooking a wire to a soldering lug, the wire should be fastened so securely to the lug that it will stay in place *before* any soldering is done.

To do this, first poke the wire through the hole in the lug. Then tighten the wire to the lug by twisting it with the soldering aid mentioned. You may find it advisable to further "clamp" the wire with your long-nose pliers, to make the joint even more firm.

With the wire firmly fastened to the lug mechanically, we are now ready to begin soldering. But what kind of solder should we use? Here is what often constitutes a real pitfall for the beginner. The acid-core solder sold in hardware stores is worse than worthless. It is a downright menace, and can ruin your equipment. Actually, acid-core solder is intended for such metals as galvanized sheet metal. It is *not* suitable for wiring electronic equipment—*ever*. Instead, always use a noncorrosive solder designed for radio work.

Furthermore, to get good results easily, it is most important to use a particular type of radio solder called 60-40. The figures mean that the solder is 60 parts tin and 40 parts lead. The greater quantity of tin results in a solder with a lower melting point than that of the fairly common 50-50 solder. The 50-50 solder is thus more difficult to use than 60-40 solder. And under no circumstances should you even attempt to use 40-60 solder. Its high melting point often results in undesirable "cold solder" joints, characterized by a pitted or grainy appearance. Solder for radio work contains an effective flux which chemically removes any small spots of oxidation you may have missed while scraping the parts.

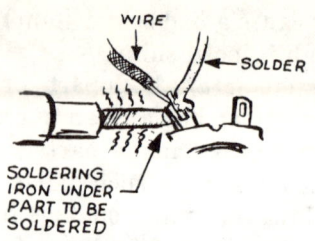

In soldering, it helps a great deal if you start off with the right idea. For example, you do *not* heat the solder with the soldering iron!

Read that last sentence again! It means exactly what it says. The proper method is to heat the *junction* (the area to be soldered) sufficiently so that the solder, when applied to the junction, quickly melts and flows onto it. Make sure the iron is hot before you start. Otherwise, the solder will pile up, and a cold-solder joint will result. Test the iron by applying solder to the tip. If it flows freely, the iron is hot. Before using the iron, wipe the melted solder off the tip with a rag.

There are two reasons for doing the job this way. First, for a good connection, all leads being joined together *must* be hot enough to melt the solder when it touches them. Second, heating the junction—instead of the solder directly—reduces the chance of burning up the flux before it has had time to do its work.

So, hold the iron underneath the joint to be made—*underneath* because solder flows like water—down. *After* the joint has had enough time to heat up, apply the solder to the *top*. As the solder melts, it should flow down around the wire, and in and out of the lug hole, until there is a bright coating over the joint. Remove the iron from the joint, and allow the solder to harden.

Examine the joint carefully. It should look shiny and smooth —not pitted or grainy. If it has the latter appearance, reheat

the joint, pry off the solder with the sharp end of the soldering aid, and start all over again.

If all looks well, give the joint time to cool. Then pull and tug on it to make certain the connection is firm. Should it break loose, count yourself lucky. You have discovered something which could cause you much trouble later. Make a new joint, and go through the same inspection again.

Like any other mechanical skill, soldering takes a little practice. But you will do all right if you will always follow these rules:

1. Use 60/40 noncorrosive radio-type solder—NEVER acid-core.

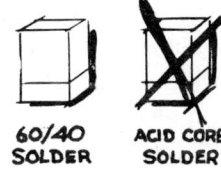

2. Scrape parts and wires clean and shiny before starting.

3. Make a good mechanical joint. Do *not* depend on the solder alone to make the mechanical bond.

4. Make sure the iron is hot.

5. Heat the joint and flow the solder on.

6. Inspect the joint carefully. If in doubt, do it again.

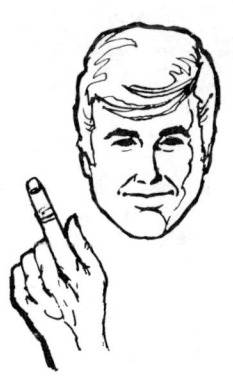

There is one other tool that is almost indispensable in some soldering jobs—the "heat sink." Although plenty of heat is needed to melt the solder and make it flow uniformly over the joint, too much heat can damage some parts such as diodes, transistors, and integrated circuits. The heat sink is clamped

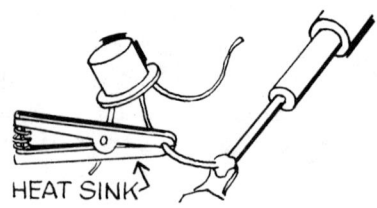

onto the lead wire between the sensitive part and the joint being soldered. The heat sink serves to soak up the excess heat that would otherwise be conducted along the lead to the part.

The photo below shows examples of the tools discussed in this chapter.

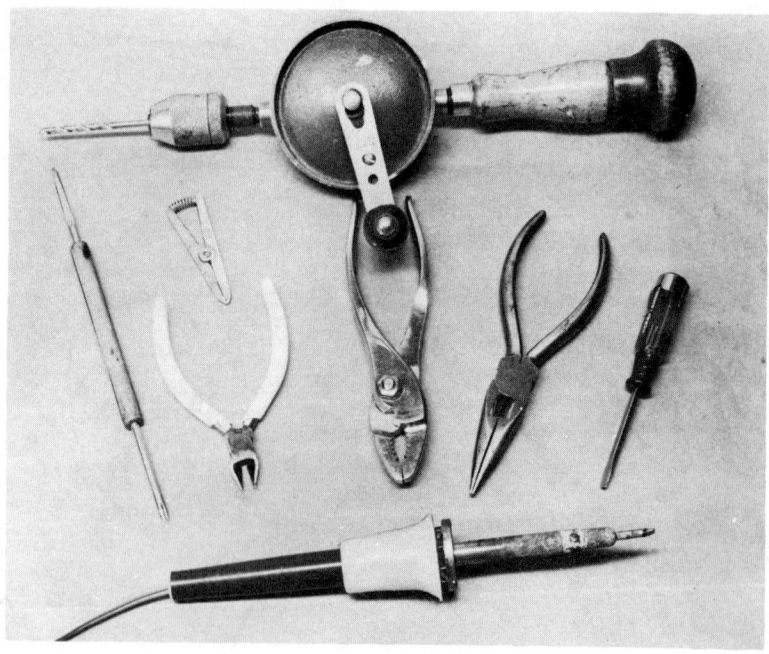

**CHAPTER 2**

# Building a Trap for Radio Waves

The fun of electronics really begins when you put together some kind of equipment which will unscramble radio waves so that you can hear them as voice or music. Doing this without some type of antenna requires a highly sensitive radio using transistors or integrated circuits that take a weak radio wave and multiply it *thousands* of times. So, the wave doesn't have to be very strong to start with; the feeble signal picked up by a "loop" within the set is good enough.

Sensitive equipment of this type is great—and some day you will want to build one yourself, even though you may already own a couple of manufactured radios. But in getting started in electronics, you'll want to tackle simple equipment until you "get the hang of things." And simple equipment is not very sensitive.

Fortunately, you can make up for this lack of sensitivity by providing an antenna and ground. In this way, you can increase the strength of the incoming signal so much that even an ultrasimple receiver (like the Two Hour radio described in the following chapter) works very well indeed. Furthermore, an antenna is a necessity for shortwave reception, which you will probably want to work with eventually. Should you decide later to become a ham operator, you *must* have an antenna in order to broadcast to other ham operators. So, before building any kind of electronic equipment, let's get started on the right foot by providing some kind of antenna.

Fortunately, erecting an antenna is a pretty easy task these days, thanks to the tv antenna manufacturers, who have worked out all sorts of brackets, masts, insulators, etc. Cost will vary, depending on the shape of your lot, height of your house, etc.

## THE INVERTED L ANTENNA

There are innumerable types of antennas, but one of the best for the beginner is a simple "L"-shaped wire. As shown in the drawings, the end of the L is brought down into the house.

As mentioned, exactly how you lay out your antenna will depend on the available space. First, we'll assume you live in a house on a typical city lot. Later on, we will consider some ideas you can use should you live in an apartment building or some other spot where there are landlords or unsympathetic people to reckon with.

An ideal antenna is one made up of 90 feet of wire. Note from the drawing that this means 90 feet from the far end at the insulator to the point where the antenna is hooked to your receiver. In a typical installation, the straightaway portion of

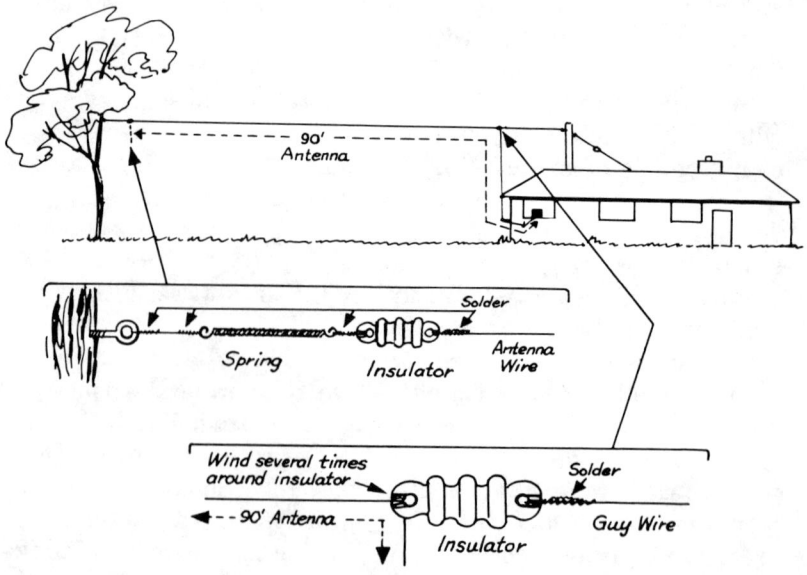

the antenna might be 60 feet, with the remaining 30 feet used as the lead-in portion. Ninety feet is an ideal length should you become a radio amateur and want a ⅜-wave antenna for 40 meters. At any rate, put up as long an antenna as you can.

Some antenna kits have a 50-foot or 75-foot length of stranded copper wire plus a length of "lead-in" wire. Make a good soldered connection between the wires.

Notice that the drawing shows insulators at both ends of the antenna. Of course, there must be something on which to hook the insulators, to support the antenna high in the air.

## TREES ARE QUITE HANDY

If your house is fairly high and there is a convenient tree in the backyard, putting up the antenna is easy. All you need to do is provide some sort of short mast on the house, and then hook the insulator (on the far end of the antenna) to the tree. Of course, since trees move around in the wind, something is necessary to take up the slack. A screen-door spring will do the job very well (see the drawings). To fasten the spring to the tree, cut a piece of wire long enough to reach between the end of the antenna insulator and the tree. Hook this wire to a husky screw eye driven into the *trunk*—not a branch—of the tree. The screw eye will not hurt the tree, and this method is much neater than the more common practice of wrapping a wire around the tree. The latter method may damage the tree.

No trees? Then perhaps a mast erected on a garage will do. An ordinary two-by-four is strong enough for up to twelve feet or so, provided it is properly guyed. The guy wires can be galvanized wire of the type used to guy tv antennas. One or more strain insulators should be inserted at least every ten

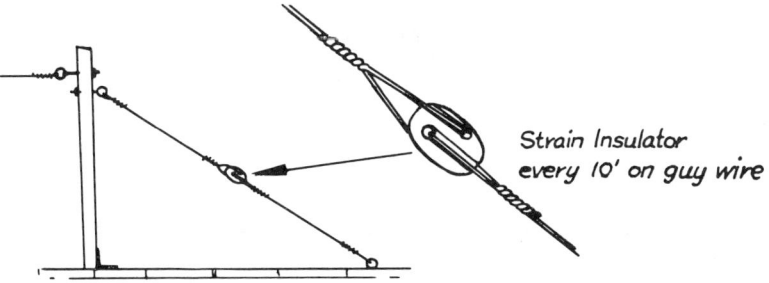

Strain Insulator every 10' on guy wire

feet in the guy wire. The idea is to ensure that no wire will be longer than ten feet—longer wires may upset reception slightly on shortwave bands. Strain insulators should be used because, if they should break, they will not open up the guy wire. The latter could be serious, depending on what is below when the antenna mast comes tumbling down!

## MOUNTING THE MAST

Erecting the two-by-four mast on the garage will, of course, depend on the type of roof, the shape of the garage, and so on. A common way to do the job is to bolt the mast to a strap hinge, and then secure the other half of the hinge to the roof

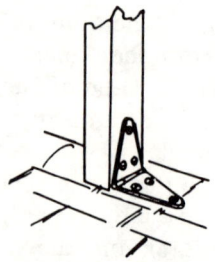

with long wood screws. Make sure the screws bite into the roof rafters or other firm structure—shingles alone are not strong enough. This is important because quite a bit of pull is involved, more than enough to lift the shingles off the roof if a heavy wind comes along and they alone are supporting the mast.

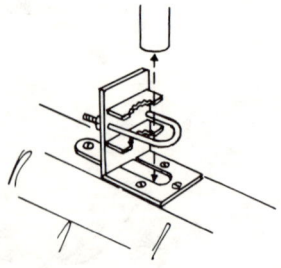

Use of tv masts provides another, and perhaps easier, way to do the job, since you can select a suitable mounting bracket from the many ones available. An inexpensive variety is adequate for the job if the tv mast is no taller than ten feet. Just as with the strap hinge, make sure the mounting bracket is screwed to something solid.

## THE WIRE ITSELF

Assuming we have worked out suitable supports for our antenna, the next step is to assemble the antenna itself. Obtain some No. 14 hard-drawn copper wire or (easier to handle) stranded bare copper wire. Size 7 × 24 stranded wire (7 strands, 24 gauge) is fine, and is one size commonly available for antennas. You will also need several tv standoff insulators

(the number depends on how many bends you have to make and how far you run the lead-in portion of the antenna). Also buy a pair of antenna insulators—inexpensive glass or porcelain insulators are good enough.

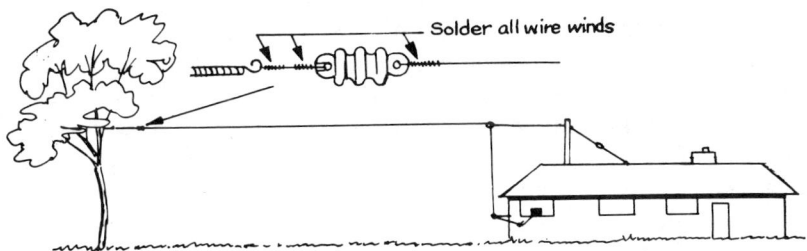

Now measure off 90 feet of wire. Fasten one end of the wire to one of the insulators by running it through the "eye" of the insulator a couple of times and then wrapping it tightly onto the wire itself. Soldering the wire is a good idea—not to make a connection, but to ensure that the joint with the insulator doesn't unravel. Run the wire to the insulator on the house. Wrap the wire through the "eye" several times, and then run it down the side of the house, leaving enough wire to go through the window (or other opening) and into the house.

How much of the wire runs horizontally and how much runs vertically is not important—just as long as you make sure that the overall length is 90 feet.

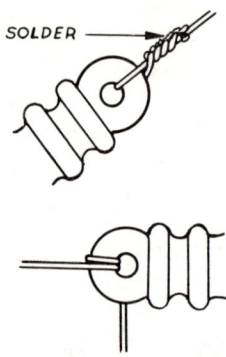

Of course, if you simply haven't the space, you can put up a random shorter length of wire, even though it will not be as effective as the longer antenna. If you must reduce the overall length, try to keep it at 45 feet—which happens to be the next shortest of the most desirable lengths.

## HOW TO OUTWIT LIGHTNING

An antenna high in the air appears to be an open invitation to lightning. Actually, the chances of a direct hit are pretty remote at the usual height for most receiving antennas. But a lightning strike anywhere in the nearby area may induce a husky current in the antenna, perhaps even enough to cause damage. For this reason, outside antennas should be equipped with a lightning arrester. (Chances are there is one on your tv antenna.) To play safe, your antenna should also have an arrester. For lowest cost, it should be the type intended for single-wire lead-in. However, these are becoming hard to get, and a tv-type lightning arrester will work just as well. Actually, the tv antenna arrester is intended for a two-wire lead-in. But no matter—we simply use one of the lead-in terminals. Nothing is hooked to the other, as shown in the drawing. The ground connection is made to a grounded rod or cold-water pipe. (See the instructions that usually come in the package with the arrester.)

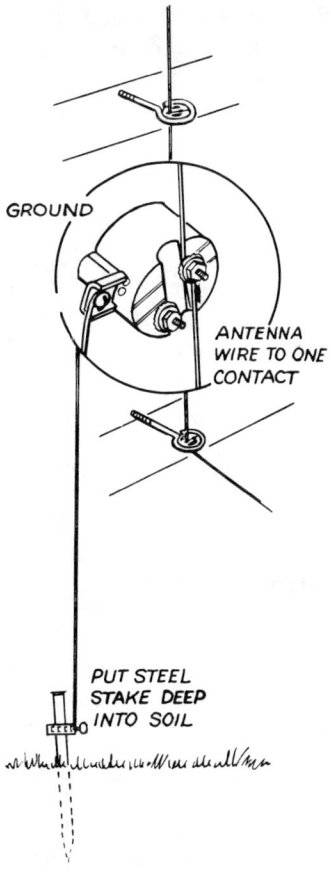

The antenna wire must be insulated from the building every foot of the way—from the insulator at the far end, right down to the antenna terminal on the receiver. Chances are you will have to run the wire alongside the building, or around the edge of a roof or some other obstacle. To keep the wire clear of such objects, the tv standoff insulators mentioned before are ideal. They come in various lengths, the 3½-inch size being very common. Screw them into the building wherever the wire must be kept clear of the building. They have a wood-screw–type point which will go into wood or, in the case of a masonry house, into the mortar joints between the bricks.

Note that the insulators have a slot in one side to permit insertion of the flat tv lead-in wire. For our purpose, we want the slot closed. To do this, simply open the slip-joint pliers as wide as possible, and then squeeze down on the wire loop around the plastic insulator. This will close up the insulator, but still leave a hole through which we can thread our antenna wire.

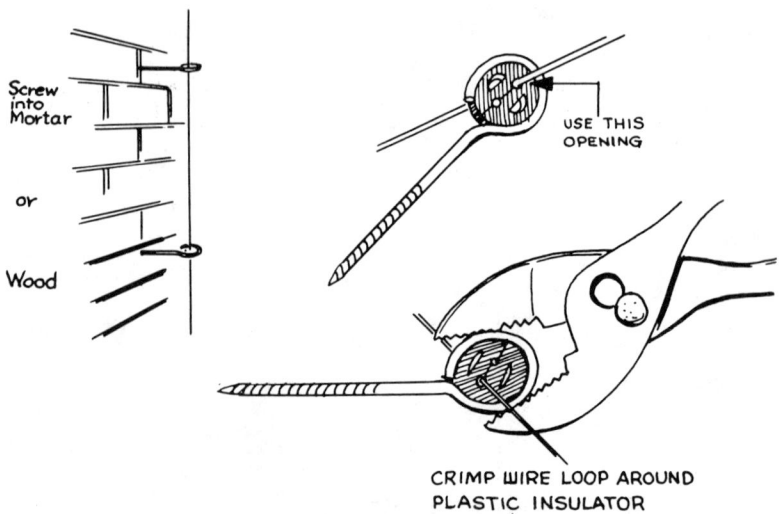

## HOUSE BREAKING MADE EASY!

As you may have suspected, getting the wire into the house through a window or other opening can be quite a chore. The best way to do the job—if you really know your way around with tools, and have a broad-minded family or landlord—is to drill a hole right through the wall. For a masonry wall, the tool to use is a star drill—a special chisel which cuts out a tiny

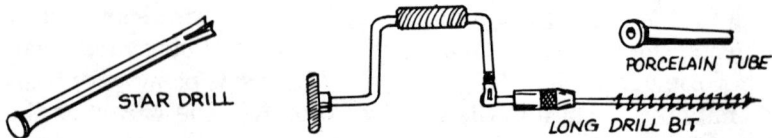

portion of the hole with each blow from a hammer. In wooden walls (usually wood siding outside, plaster inside) a long drill bit of the type used by electricians will do the job easily. Another approach, if you have basement windows with wooden casings, is to drill through a casing, and then up through the floor, under the quarter-round trim. Once you have the hole through, you can slide in a tubular porcelain feed-through insulator—or, for better appearance (but higher cost), one of the tv-type through-the-wall insulators.

Somebody in your family takes a dim view of holes in the house? Don't despair; you have lots of company—and there are less drastic ways to do the job.

Perhaps the simplest way is to run the antenna wire between the window and the sill. Of course the wire must be insulated *from* the window. This is particularly important with the metal windows so common today.

One way to do this is to use a short length of the high-voltage plastic tubing commonly used to protect the high-voltage leads in tv sets. Any good tv shop should be willing to sell you

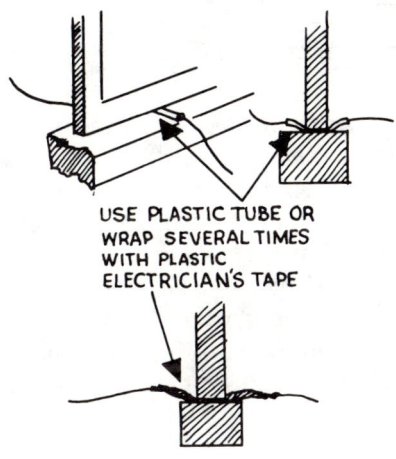

USE PLASTIC TUBE OR WRAP SEVERAL TIMES WITH PLASTIC ELECTRICIAN'S TAPE

a foot or so for a few cents. If you can't find the tubing, wrapping the wire with half a dozen layers of plastic electrician's tape will insulate it fairly well. Once the wire is insulated, simply close the window on it, being *careful* not to scrape or cut a hole in the insulation.

## LOOK, MOM—NO CONNECTIONS!

Yes, there is even a way to get through a window which can't be opened. The answer is to make a kind of sandwich, with aluminum foil as the "bread" and the window glass as the "ham" in between.

To make this metal sandwich, bare the end of the antenna wire that comes down the side of the house to the window. Then take a one-by-two foot (or larger) piece of aluminum foil and fold it over, being certain it makes good contact with the wire. (See drawings.) Cement the foil to the outside of the window with Duco china cement. (This type of connector can be purchased if you don't want to make one.)

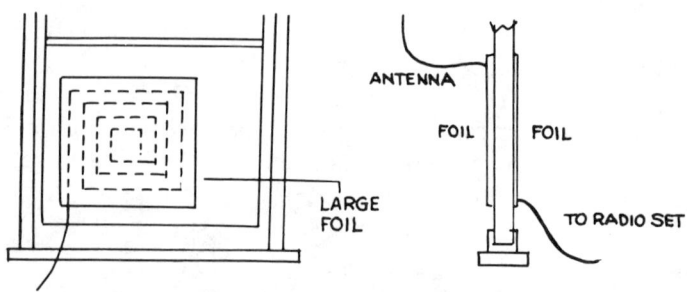

Repeat the process inside, providing a lead-in wire which can be connected to the square aluminum sheet (double thickness) and to the antenna connection on the receiver. There is NO direct connection between the two pieces of foil—the glass is between them. What we have done is create a large *capacitor* (often called a *condenser*)—the pieces of foil on both sides form the *plates*, and the glass in between forms the *dielectric*. High-frequency radio waves will pass through such a capacitor without serious loss. The larger the pieces of foil, the less the loss.

## FOR APARTMENT DWELLERS ONLY

An outside antenna is the best thing to use. The shortcuts which follow are intended *only* as a last resort, but if you live in an apartment or college dormitory, they may be the only way out.

One solution is to use an indoor antenna made up of flexible wire. If you are on the second or third story of a building, the wire may do a passable job if strung along the outside edge of the floor. If you can sneak it down the hall, still better; every inch counts. The top edge of the wall of the room (see draw-

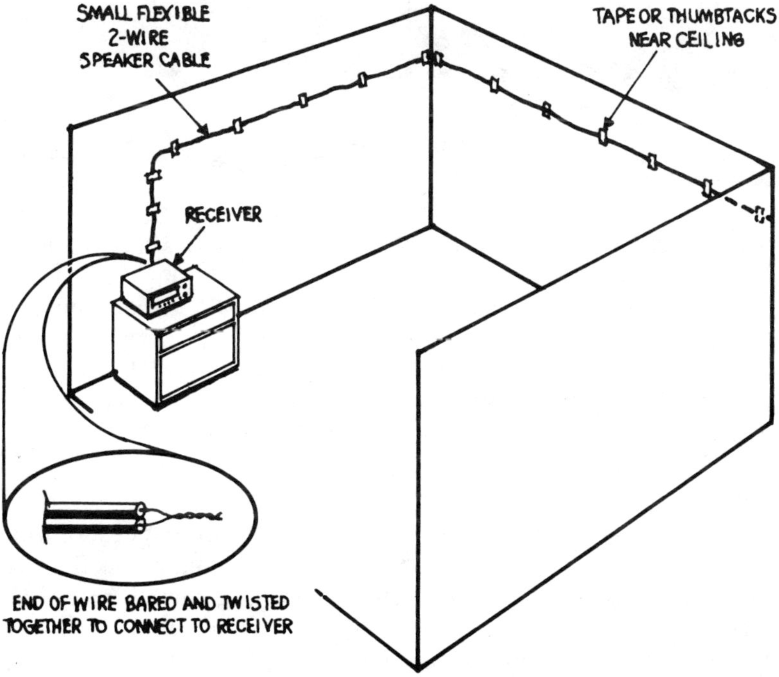

**INDOOR ANTENNA**

ing) may be used. At any rate, get out all the wire you can—don't worry about the exact length.

*Sometimes* simply running a wire and a clip to the finger stop on a phone will provide a usable antenna. A somewhat

better solution is to "capacity couple" to the telephone line, and "hitchhike" on it as an antenna. One way to do this without disturbing the telephone is to wrap at least three feet of the flexible antenna wire around the telephone cord coming

from the wall. The result is a combination of inductive and capacitive coupling, but *no* direct connection to the telephone line.

How well this makeshift antenna works will depend on the amount of metal in the building, particularly in the conduit system (if any). However, it often works surprisingly well.

Other possibilities include tapping onto an outside fire escape, downspout, or other metal on the outside of the house. Whether or not they will work can only be determined by experimenting with them.

Here is a final suggestion, if you have an understanding family and an *outside* tv antenna. The tv antenna itself isn't big enough for our purpose, but by hooking onto the lead-in line, we can use it for our antenna. One way to do this is to run a short wire from the receiver we are going to build, to an alligator clip. The latter is simply clipped to the antenna terminals on the tv set whenever we want the tv antenna to fur-

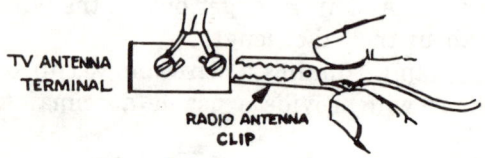

nish signals to our small set. The tv set can remain connected at all times. But when you want a good picture on the tv set, you'll have to remove the clip!

In addition to an antenna, we need a ground. It can be an outdoor ground like the one mentioned previously, provided we can get a wire out to it. Or, start over again by running a wire (either bare or insulated) to a clamp on a *cold-water* pipe inside the house. Hot-water pipes are not as effective because the hot-water tank is between the connection and the cold pipe. Gas pipes are very poor. Furnace outlets or steam pipes often work very well.

Another possibility for a ground is the mounting screw on a wall receptacle. In most houses or buildings the metal box is grounded. See the illustration.

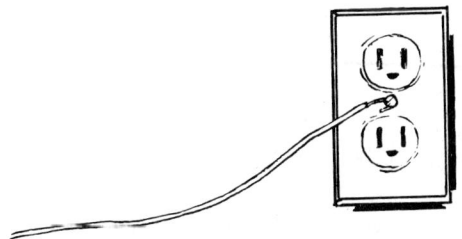

So, we have an antenna and ground ready to be connected to our first radio. The antenna is fairly jumping with all sorts of radio signals. And it is surprising how little is needed to transform them into something we can hear. That is exactly what we will do next, by building the Two Hour radio described in the following chapter.

CHAPTER 3

# The Two Hour Radio

Learning about electronics without *building* electronic equipment is like learning to swim without getting into the water. So let's dive in and build something!

Our first project is a radio with practically no parts at all. Yet it will tune in strong *local* broadcasting stations in fine style. And late at night, with a good antenna like the 90-foot inverted L described in the previous chapter, don't be surprised if it pulls in stations up to 200 miles away (particularly after local stations go off the air, giving the weaker out-of-town stations a chance to be heard).

This receiver uses a kind of crystal detector called a *germanium diode*. This is a modern device which will handle far stronger radio signals than the old crystal did—and it is also more sensitive. When combined with another part, a homemade coil (an hour's job), the result is a real radio capable of practical reception despite its simplicity. Because the parts are few, you can actually wire the set in another hour—hence the name, *Two Hour* radio.

## HOW IT WORKS

Explaining in a few paragraphs how a radio works is quite a trick—many shelves of books have been written on the subject. But before wiring up our set, let's take a moment to consider what our little radio must do if we are to hear anything from it.

Sometimes the easiest way to understand anything complicated is to compare it with something familiar. In this case, we will look at the telephone.

First of all, we need to remember a basic fact about electricity. Electrical current is a flow of electrons that behaves something like a flow of water. For example, if water is flowing in a garden hose and you step on the hose, the water flow will be *reduced*. Electrical current works in exactly the same way

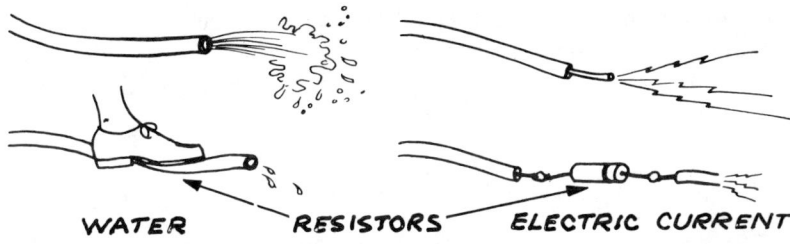
WATER — RESISTORS — ELECTRIC CURRENT

when it moves along a wire. If we add *resistance* (just as we added *resistance* to the water flow when we stepped on the garden hose), we will *reduce* the flow of electrical current. (That is what a *resistor* does in a radio.)

Now—as you may already know—the microphone in a telephone is actually a kind of *variable* resistor which affects the flow of current in the telephone line. When you speak into the microphone, sound waves produced by your voice strike a thin metal plate (diaphragm) inside the "mike." As the plate moves in and out, it alternately compresses and releases the pressure on the carbon particles inside the microphone. This changes the *resistance* these carbon particles offer to the current. Thus, as you speak, you vary the amount of current through the microphone and the telephone wire, which is connected to a receiver at the other end of the line.

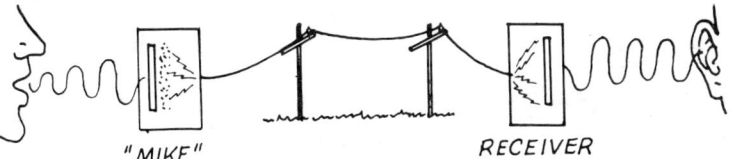
"MIKE"            RECEIVER

The receiver you hold to your ear *responds* to the variations in current, *exactly in step* with the words being spoken into the microphone. The diaphragm in the receiver goes in and out, also in step with the movement of the diaphragm in the microphone. As the receiver diaphragm moves, it re-creates the sound waves produced by the voice at the other end of the line.

With a telephone, this process is easy because *a wire* connects everything together. But what about radio (or "wireless," as it was originally known)?

Remember that in our telephone we carried sound waves from one place to another by first causing the pressure of the sound waves to vary the current in the telephone wire. At the receiving end, we transformed this varying current back into sound waves. The *wire* simply serves as the vehicle which carries our varying current. The current *variation* is the important item.

## TO THE MOON AND BACK

If, in some fashion, we could use radio waves to carry the effect of our varying current, we would have a way to span continents—or for that matter, to carry a voice to the moon and back. Of course, this is exactly what does happen in radio. We "hitchhike" on radio waves sent out by a broadcasting station. At the broadcasting station, we speak into a microphone and thus set up a varying current, exactly as we did on the telephone. In fact, you may have heard telephone conversations being broadcast over the radio. By a process known as "modulation," the varying current from the microphone is combined with the radio wave in such a way that the *amplitude* of the radio wave varies with that of the voice or other sound.

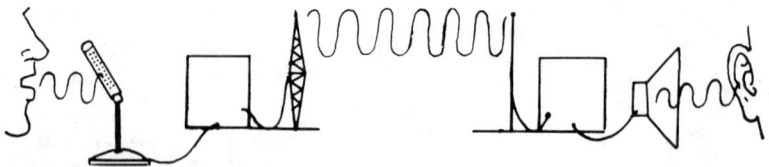

Our radio wave, complete with sound modulation, may travel to a radio set thousands of miles away. But, to get something we can hear, we must translate our incoming signal back into the form it had before it was combined with the radio wave.

Fortunately, this is easy to do. At the receiving end, we simply feed our modulated radio wave into one end of a germanium diode, the modern version of the crystal in the old-fashioned crystal set. From the opposite end of the diode comes the same kind of current we had on the telephone line. So, we have only to apply this varying current to a headphone or speaker, and we will hear sound.

Yes, it is possible to build a radio with only a diode, some kind of earphone, and an antenna and ground. Such a radio

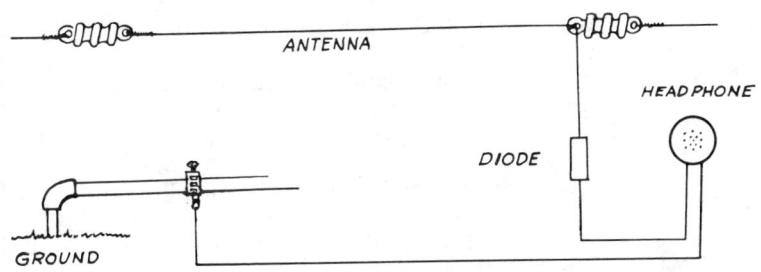

will receive signals after a fashion, particularly if there is a strong station nearby. If there are two strong stations nearby, chances are you will receive *both* at the same time, which is pretty discouraging.

One important element is missing: some type of tuning device to enable us to choose the radio waves we want. The most common way of doing this is to use a coil and a variable capacitor. By properly combining these two parts, we can be more selective in our choice of stations, tuning in one and tuning out others so that we hear them only weakly or not at all.

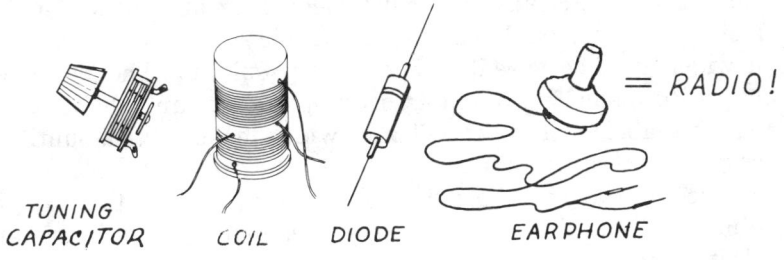

A simple radio, then, needs a tuning capacitor, a coil, a germanium diode, and an earphone. That is exactly what we use in the Two Hour radio.

## Shopping List

| Quantity | Description | Label in Drawing |
|---|---|---|
| 1 | Perforated board, approximately 4½ × 5¾ inches (11.4 × 14.6 cm) with 0.093-inch holes. Overall size not critical for this project. | |
| 1 | Variable capacitor, 365 pF with dial (Calectro A1-232 or equivalent) | C |
| 1 | Subminiature phone jack | J |
| 5* | Fahnestock clips | |
| 1 | Pill bottle, clear plastic, 2 inch (5.1 cm) diameter (see text) | |
| 6* | 6-32 machine screws, ½ inch (1.27 cm) or longer, with nuts | |
| 7* | Soldering lugs | |
| 1 | 1N34A diode or equivalent | H |
| 2 | Cleats (see text) | |
| 4 | Screws for cleats (see text) | |
| 1 | Earphone (see text) | |
| | Magnet wire, No. 28 (or No. 26 or 30—see text) | |
| | Insulated hookup wire, No. 22 | |

*One more needed for T-7 (see text).

## LET'S START BUILDING

In this first set, we will use a modern "breadboard" method in which the parts are assembled on a perforated "board"—a thin plastic material which is punched with many holes. This is a standard item available from electronic supply houses, and it is widely used because it makes for easy building. The material has good insulating qualities, is simple to cut and shape, and has available various little fittings which make parts mounting easy.

Select a board of approximately the size given in the "shopping list," and then mount it on some small wooden or metal cleats. The cleats are necessary to raise the board off the oper-

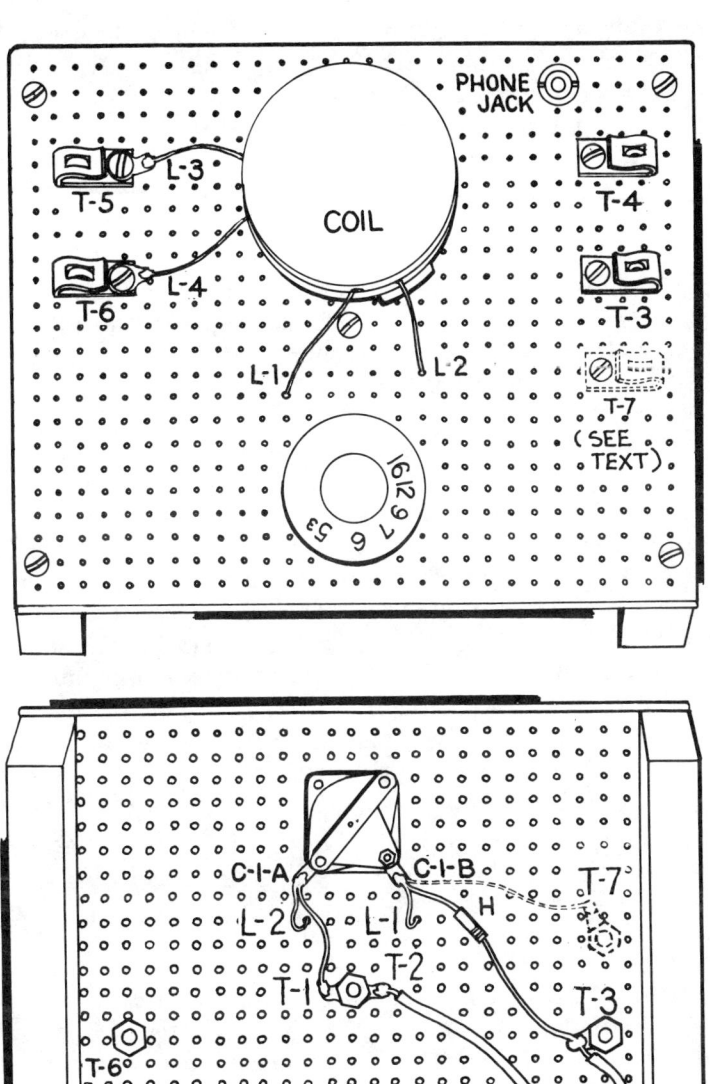

PICTORIAL DRAWINGS

ating table or surface so that there is room underneath for parts. These cleats should be approximately ¾ inch square. Small rubber feet are another possibility.

The first step is to enlarge four of the holes in the board so that they are large enough to allow attaching the board to the cleats with small screws. The pictorial drawings show how this is done, using four screws.

Since this is your first building project, we will begin by using a step-by-step construction procedure. Follow the instructions as given, and check off each step in the little box provided. If you will do this systematically, you will be far less likely to make a mistake.

- ☐ With wood screws, fasten the board to the small wooden cleats. (Metal brackets could, of course, be used instead if they are more readily available.)
- ☐ In the fifth row of small holes from the front, drill a ¼-inch hole approximately centered from left to right. This hole is for mounting the variable capacitor.
- ☐ The next step is drilling a hole for mounting the subminiature phone jack. (This jack and the wiring which connects to it are not required if the earphone you will be using has two leads which can be connected to terminal clips T-3 and T-4.) Jacks of this type normally require a 3/16-inch hole, but it is a good idea to check the drill bit against the jack itself to see if a hole of that size will provide the proper clearance for easy mounting.

Before going any further, we will take time off to wind the tuning coil.

This coil is homemade and is wound on a clear plastic pill bottle obtainable from a pharmacy. These bottles are approximately 2 inches in diameter and come equipped with a plastic lid.

Ideally, the coil should be wound with No. 28 magnet wire. However, either No. 30 or No. 26 wire can be used instead.

**Rolling Your Own Coil**

Starting at the top of the bottle, but far enough to clear the lid, drill two small holes directly opposite each other. This can be done with a small drill bit, or with a small finishing nail

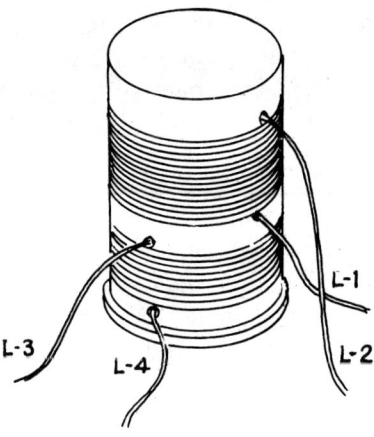

"chucked" in the hand drill. If you use a nail, it will work best if you first clip or saw off the head.

Push the wire through the two holes so that the wire will "lock" onto the pill bottle "coil form." Leave a piece about 6 inches long extending from the form.

Now, to wind, rotate the coil form with one hand, and guide the wire with the other, adding the turns as close together as possible without their climbing over each other. Wind on 30 turns. At this point, drill two more holes, and push the wire through, leaving a piece about 6 inches long for making connections later.

Repeat the process to make the larger coil, which should have 70 turns.

Don't be concerned if you lose count in winding the coil, especially if you wind on a few extra turns for good measure. The easiest way to count the turns is to run a small screwdriver down the coil "feeling" the wires with your forefinger as the screwdriver climbs over each wire. Do this a couple of times and strike an average if you are not positive you have the right number. Fortunately, the number of turns on either coil is not particularly critical.

Note that two of the coil leads come to the side where they hook to the antenna terminals, while the other two from the larger coil go down beneath the board and hook into other parts of the circuit.

Coil wound? We are now ready to start mounting parts again.

- [ ] Drill a small hole in the plastic bottle lid, and, using this as a template, drill a hole in the perforated-board base. Then bolt the lid to the back of the panel.
- [ ] Directly in front of the bolted-down lid, drill a hole for a 6-32 bolt (machine screw), and mount two soldering lugs, T-1 and T2, on the underside of the board. These form a terminal point.
- [ ] Mount the earphone jack.
- [ ] Mount the 365-pF variable capacitor to the board, using a suitable mounting hole. If you are using a capacitor like that shown, you will find that the metal center of the dial unscrews and allows the plastic dial to be removed. After this is done, you can remove the nut that is used to secure the variable capacitor to the board. The nut on the shaft is simply tightened down to secure the capacitor on the board. Do this with a pair of pliers, but do it gently so that you do not strip the threads on the shaft or crack the board.
- [ ] The next step is to mount Fahnestock clips T-3 and T-4 along the edge of the panel close to the phone jack. Note that the clips are both mounted with bolts, and underneath the panel are nuts and soldering lugs.
- [ ] In similar fashion, Fahnestock clips T-5 and T-6 are mounted on the other side of the board, except this time the terminal lugs are mounted above the baseboard to allow connecting to the coil.

We are now ready to start soldering.

**Don't Forget the Proper Soldering Technique**

Start with a good, hot soldering iron. The iron should be wiped clean with a damp cloth. If it has not already been

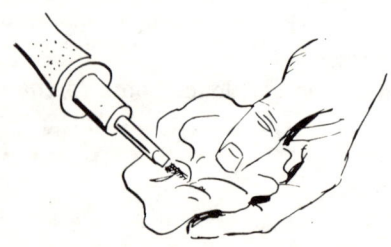

"tinned," sand it slightly (or file, ever so lightly), and, with the iron hot, melt some solder on the tip surfaces.

As mentioned previously, in making connections, secure the parts and the wire leads together mechanically, and *then* flow on the solder. Don't depend on the solder simply to stick things together. Finally, be certain that you are using 60-40 solder. *Do not use acid-core solder under any circumstances;* it will absolutely ruin any electronic gear.

First, we'll hook up the coil. Note that the leads on the coil are marked L-1 through L-4. We'll start with lead L-4 at the bottom end of the coil, and this goes to T-6 as shown on the drawing. The enamel insulation on the wire must be removed. This can be done either by scraping it very carefully with a pocket knife, or—and the writer has always found this easier —with a piece of fine sandpaper. At any rate, be certain that the wire is *clean* of enamel. Otherwise, solder will not stick to it.

- ☐ Solder L-4 to T-6.
- ☐ Solder L-3 to T-5. This completes the wiring above the baseboard. Turn the baseboard upside down, and we will complete the wiring.

From here on, you will see the notations (S) and (DS). If you read (S), it means make a soldered connection. If you find (DS), it means make a mechanical connection. Frequently this means bending a wire through a hole or terminal point, but do not solder yet because another wire will be connected to that same point later, and then both will be soldered.

- ☐ Connect coil lead L-2 to terminal C-1-A on the variable capacitor. Note that this connection is made to the plates of the variable capacitor which do not turn (called the "stator" of the capacitor). The other terminal point on the capacitor goes to the bearing and shaft of the capacitor, thus providing an electrical connection to the rotating portion of the capacitor (called the "rotor") (DS).
- ☐ Connect L-1 to C-1-B (DS).

Now we will wire our first part, in this case the diode, which is labeled *H* in the diagram. This small diode, a 1N34A or equivalent, will be marked on one end with a band (or two), and it should be wired into position with the band as shown, which in this case is on the end opposite that which connects to the variable capacitor.

☐ In connecting the diode to capacitor terminal C-1-B, make a mechanical connection first, and then use pliers or a heat sink to protect the diode from heat while you do the soldering job (S). If you are using a variable

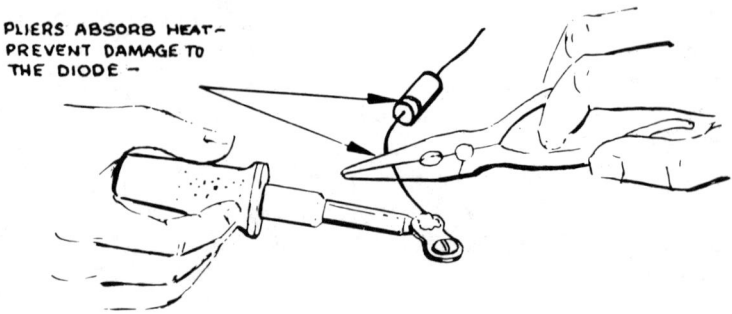

PLIERS ABSORB HEAT—
PREVENT DAMAGE TO
THE DIODE —

capacitor like that shown with plastic between the plates, use the *minimum* heat necessary to solder C-1-B. If you overheat C-1-B. you may melt the plastic and ruin the capacitor. See Chapter 4 for more details.

After you have made the connection, pull and tug on the diode to be certain that the connection is a good one. Examine the connection carefully. Remember, *more home-built electronic equipment fails to operate because of bad soldered connections than for any other reason!*

☐ Run the other end of the diode to terminal T-3 (DS). Again protect the diode with pliers. Wire a short lead from T-3 (S) to lug J-3 on the phone jack (S). If you are using plastic-covered wire, the insulation can be stripped off with your side-cutting pliers, providing you do the job carefully, as described in Chapter 1. However, a very useful tool to add to your tool kit is a wire stripper like that shown in the photo.

☐ Bare approximately 1 to 1½ inches of a piece of hookup wire, and run it through terminal T-4 (DS) to J-1 on the phone jack (S).

☐ From T-4 (S), run a lead to T-2 (S).

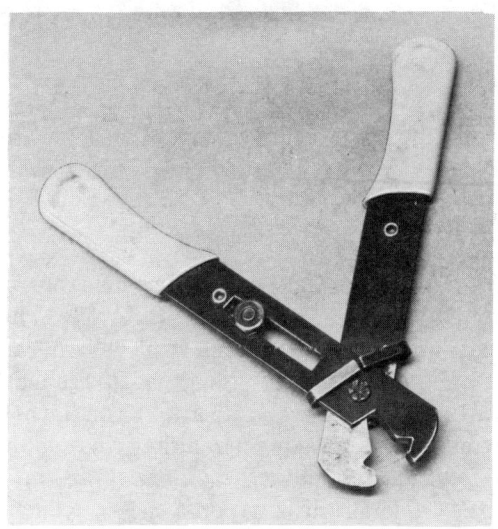

☐ Connect a lead from T-1 (S) to C-1-A (S). This completes the wiring.

Recheck the wiring sequence step by step. Check the wiring against the drawings. If everything is okay, we are ready to give our little set its first tryout.

Connect whatever antenna is available to T-5. Connect the "ground" to T-6. Hook up the crystal earphone either to T-3

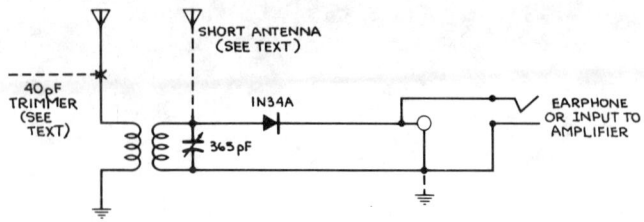

and T-4 or to the phone jack, depending on what type of cord end is on the earphone. *MOST IMPORTANT:* This earphone must be of the "crystal" type, which is of very high impedance. A low-impedance earphone of the type furnished with small transistor radios will not work at all on the Two Hour radio. A 2000-ohm or higher-impedance magnetic earphone will function after a fashion, but not *nearly* as well as the crystal earphone, which is also far less expensive.

Earphone hooked up? Rotate the variable capacitor. If there is an am (standard broadcast band) radio station within range, you should pick it up. In Denver, where the writer lives, the Two Hour radio will pull in ten different stations when a good outside antenna is used. Even with the "tv set" antenna described in Chapter 2, six stations come in with satisfactory volume.

If you have a very poor antenna, and there are no stations within 10 miles or so, don't despair. Chapter 5 tells how to increase volume a great deal even with a poor antenna, and subsequent chapters tell how to build far more sensitive sets as well.

Perhaps you are in a city like Denver and have the opposite problem: lots of stations, and some of them coming in on top of each other. You can improve the selectivity of the set—with some loss of volume—by connecting a trimmer capacitor in series with the antenna. A better approach is to build the unit in Chapter 4, which has a few more parts, but is also far more selective, since it has two tuned circuits instead of only one.

## POOR ANTENNA, WEAK SIGNALS?

Should you have only a makeshift antenna available and be quite a distance from the transmitters of the local broadcasting stations, there is one more thing you can do.

First of all, get the *best* ground connection you can; if possible try several *different* ones (one at a time) : water pipe, connection to the electrical outlet box, etc. Hook the ground lead to T-4, along with the earphone.

Next, modify the set slightly. The circuit uses "inductive coupling." The current in the antenna coil "induces" current into the tuning coil; there is no direct connection. The advantage of this approach is that it provides better selectivity when there is only one tuned circuit. However, there is some signal loss as well. To overcome this problem in a weak-signal area, we can connect the antenna directly to the tuning coil by providing another Fahnestock clip and a lead running directly to the tuned circuit. The dash lines in the pictorial diagrams and circuit diagram show how to do this.

When we connect the antenna to the tuned circuit we "load" it and require less coil to tune a given frequency. Unfortunately, there is no way to predict how much loading will occur; that depends on the antenna being used. However, if a station that should be at about 550 on the dial comes in at 600, there is too much coil. So, remove turns from the top end of the coil five turns at a time until the station appears about where it should on the dial.

**CHAPTER 4**

# Selective AM Tuner

The ability of any kind of am radio to separate stations from one another depends a great deal on how many tuned circuits there are in the hookup. In the ordinary am radio of the superhet variety, there are frequently as many as six tuned circuits, counting those in the if amplifier stages.

A superheterodyne am receiver is a bit difficult for the beginner in electronics. It is better to buy a good ready-made unit! However, we can greatly improve the selectivity of a simple diode radio, like the one described in the previous chapter, by using not one but two tuned circuits. The objective of

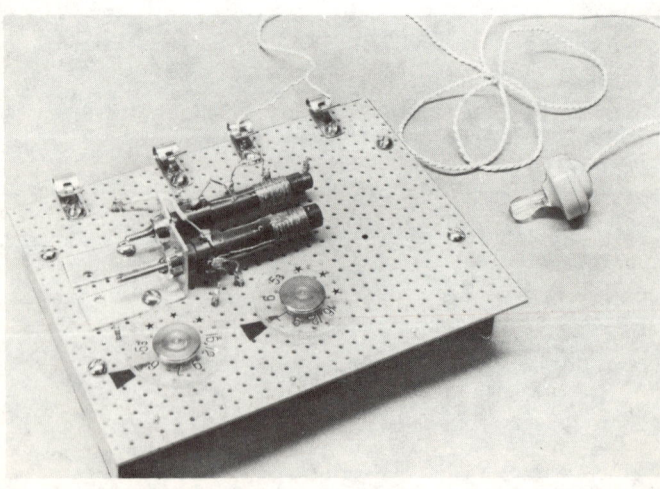

increasing selectivity can also be helped along by using coils that have powdered-iron cores, which provide a characteristic called "high Q."

## Shopping List

| Quantity | Description | Label in Drawing |
|---|---|---|
| 1 | Perforated board, approximately 4½ × 5¾ inches (11.4 × 14.6 cm) with 0.093-inch holes | |
| 11 | Push-in terminals for 0.093-inch holes | |
| 4* | Fahnestock clips | FH |
| 4 | Solder lugs | T |
| 6* | Machine screws (6-32, ½ inch long) with nuts | |
| 2 | Cleats (see text) | |
| 4 | Screws or bolts (to suit cleats used) | |
| 2 | Variable capacitors, 365 pF with dial (Calectro A1-232 or equivalent) | C |
| 2 | Ferrite-core antennas (Calectro D1-841 or equivalent) | L |
| 1 | 1N34A diode or equivalent | D |
| 1 | Earphone, crystal type | |
| | Hookup wire, No. 22 | |

*One more needed for optional antenna terminal (see text).

The photograph shows a simple radio which has two tuned circuits. You will note that it utilizes some of the parts from the Two Hour radio, and the building technique is much the same. As before, we start out with a piece of perforated board mounted on wooden strips, metal cleats, or rubber feet.

## A TECHNIQUE TO LEARN

To help with mounting the parts and duplicating wiring, we will use a technique that will be repeated in subsequent chapters. The idea is to identify each hole on the perforated board by combining letters and numbers. The top view drawing shows how this is done. Note that the holes are numbered across the board and letters are used to identify the holes along the side of the board.

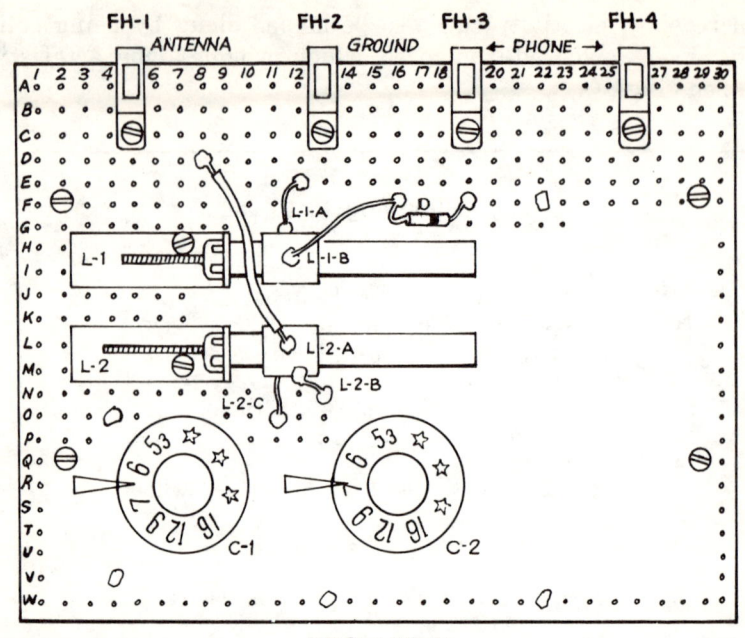

TOP VIEW

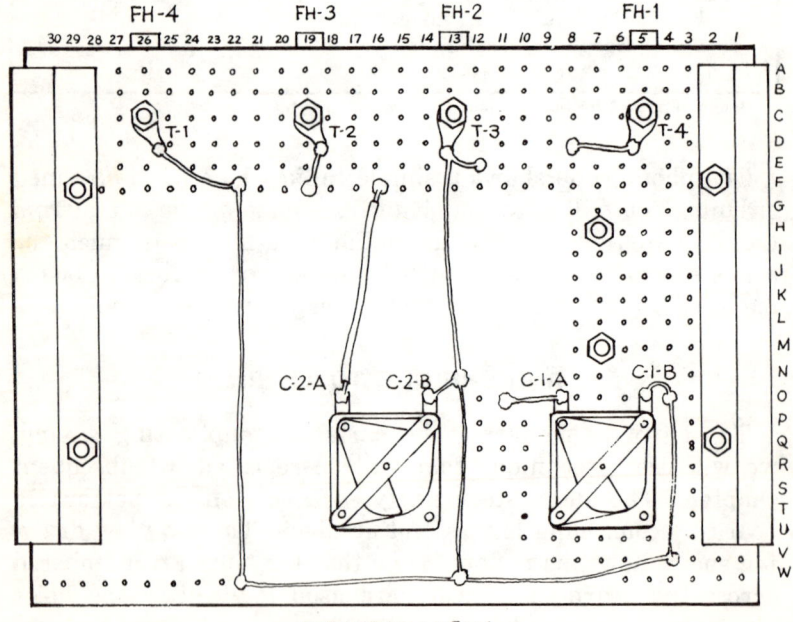

BOTTOM VIEW

Start out by writing the numbers and letters directly on the board with a soft pencil, starting on top of the board. No, there is *no* need to mark each one of the holes! If you have the numbers at the top and the letters at the lefthand side, it is a simple matter to run your eye across a row and identify a given hole.

After you have prepared the top side of the board, turn it over and go through the numbering process again. But note that *this* time the numbers go from right to left. The letters remain the same, of course, on either the top or bottom of the board.

In a set of this type, parts layout is not at all critical, and considerable variation is possible without effecting results. However, we might as well start out and do things in an orderly way. In some of the units described later in the book, layout *is* quite important.

### INSERTING CLIPS

With our board lettered and numbered, we are ready to start. The first step is to insert the push-in terminals into the board. These small metal parts come by various names, such as "Flea Clips" and "Zip" terminals. By whatever name, they should be the size to fit the 0.093-inch holes in the board.

There are tools made for pushing the clips into the board, but a perfectly satisfactory way to do the job is to clip one lead on a 2-watt resistor so that approximately ¼ inch remains; then use this to shove the clip into the board (see illustration). Insert clips at the following locations; all clips are pushed in from the top.

| *Row* | *Holes* |
|---|---|
| W | 13, 22 |
| V | 4 |
| O | 4, 11 |
| N | 13 |
| F | 16, 19, 22 |
| E | 12 |
| D | 8 |

Next, mount Fahnestock clips and soldering-lug terminals at holes C-5, C-13, C-19, and C-26; enlarge the holes with a

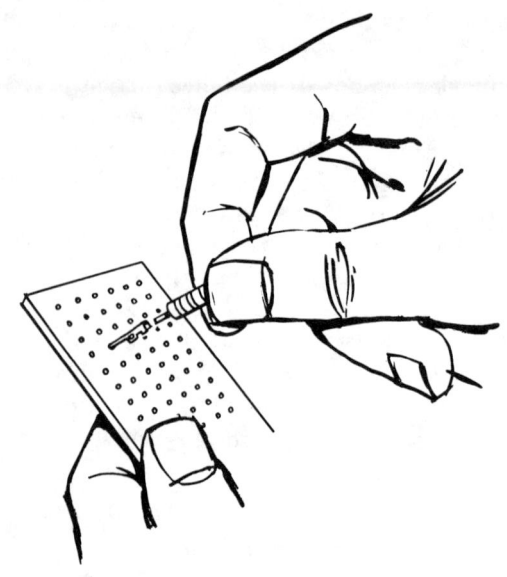

$5/32$-inch drill bit. Use 6-32 machine screws (bolts) and nuts. The bolts go through the clips, with soldering lugs T-1, T-2, T-3, and T-4 underneath the board.

The two variable capacitors are mounted in enlarged holes R-7 and R-16.

The ferrite-core-antenna coils usually come equipped with metal brackets bent at right angles as shown in the illustrations. The coils are pushed into large holes in the brackets and "snapped" into place. The brackets are secured to the board with bolts and nuts.

## WIRING

Now we will do the wiring on the underside of the board. Use No. 22 solid wire. You can strip all the insulation off the wire so that it is bare, or bare only the ends if you prefer. Both techniques were used with the model illustrated.

*Before starting,* read again the soldering instructions in Chapter 1. *Remember,* more homemade equipment fails to operate because of bad soldered connections than for any other reason. The second cause of trouble is wiring mistakes. Defective parts are a very distant third.

In some cases, as the first move we will simply make a mechanical connection of the wire to the terminal, clip, etc. (by bending the wire through a hole, or whatever else is indicated). When we make only a mechanical connection, the notation will be DS (Don't Solder). This is to allow adding one or more parts later, at which time the soldering is completed (S).

Make these connections:

- ☐ T-1 (S) to F-22 (DS)
- ☐ F-22 (S) to W-22 (DS)
- ☐ W-22 (S) to W-13 (DS)
- ☐ W-13 (DS) to V-4 (DS)
- ☐ V-4 (S) to O-4 (DS)
- ☐ W-13 (S) to N-13 (DS)
- ☐ N-13 (DS) to T-3 (DS)
- ☐ T-3 (S) to E-12 (S)
- ☐ T-4 (S) to D-8 (S)
- ☐ F-19 (S) to T-2 (S)

At this point the soldering becomes a bit more delicate, since we will be making connections to the variable capacitors. These units have thin plastic sheets *between* the plates, and these sheets are very easily damaged if overheated. The way to avoid problems is to bend the end of the hookup wire around the terminal point (for example, C-2-A) to achieve a mechanical joint, and then use long-nose pliers or a "heat sink" tool (see Chapter 1) to prevent the heat of soldering from reaching the capacitor plates. Connect

- ☐ C-2-A (S) to F-16 (S)
- ☐ C-2-B (S) to N-13 (S)
- ☐ C-1-A (S) to O-11 (S)
- ☐ C-1-B (S) to O-4 (S)

This completes the wiring under the board. Turn the board over, and we will wire on the top side.

Next, wire in the coils as shown in the top view drawing and the coil connection diagram.

- ☐ L-1-A (S) to E-12 (S)
- ☐ L-1-B (S) to F-16 (DS)

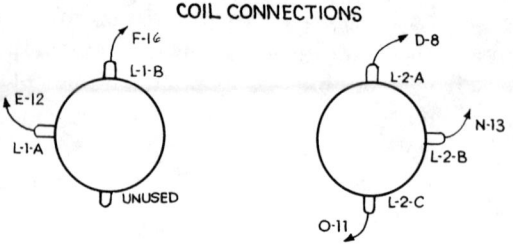

**COIL CONNECTIONS**

☐ L-2-C (S) to O-11 (S)
☐ L-2-B (S) to N-13 (S)
☐ L-2-A (S) to D-8 (S)

The only part left is diode D. Solder it between F-16 (S) and F-19 (S). Use the method shown in the previous chapter to protect the diode from heat when making connection to either end. Note that the "banded" end of the diode goes to F-19.

As shown in the illustrations, there are small, black pointers under the dials. The pointers are simply cut from electrician's tape and pressed down on the board under the dials.

**TESTING**

Connect the antenna to FH-1. Unless you live within a couple of blocks of a broadcasting station (the *transmitter*, not the studio!) you will need an antenna, preferably outside.

Connect a ground to FH-2. This can be a water-pipe ground, a connection to an outlet box as explained previously, or, in some cases, a metal heating duct or radiator pipe.

The actual tuning is done with C-2. Set this to one of the stronger stations in the area (it should come in with approximately the same dial setting as on any am radio). Now rotate C-1. This should "peak up" the signal.

If the two dials read approximately the same, no further adjustment will be needed. Probably they will not agree—coil L-2, being loaded down by the antenna, will have too much inductance. Fortunately, the remedy is simple:

1. Set C-2 to the strongest signal you can pick up toward the 53 end of the dial (C-2 almost closed).
2. Set C-1 to the same dial reading.

3. With a screwdriver, turn the tuning slug on L-2 until the incoming signal is the strongest.

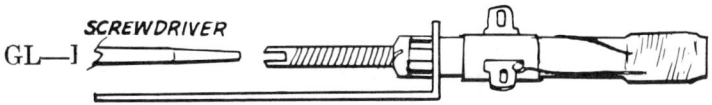

Now the dials should read approximately the same when a station is tuned in with C-2 and peaked up with C-1.

Unless you have quite a long outside antenna, selectivity will probably be satisfactory. If not, antenna coupling can be reduced (with a loss in volume) by inserting a trimmer capacitor in series with the antenna, as described in Chapter 3.

## WANT TO USE A LOUDSPEAKER?

Tired of listening on an earphone? The next chapter describes an integrated circuit amplifier which will increase the volume enough to operate a loudspeaker with ease on any station which gives good earphone volume.

## POOR ANTENNA?

If you have a poor antenna, just as with the Two Hour radio you can increase signal pickup by connecting whatever antenna you have directly to the first tuning coil. This is shown by dash lines in the circuit diagram.

To make the connection, run a wire lead from an additional Fahnestock clip to point O-11. Adjust the slug in L-2 so that the tuning capacitors match up and the dials read properly.

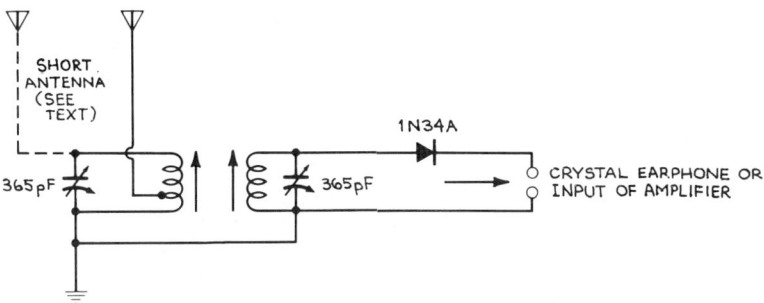

# CHAPTER 5

# Loudspeaker Amplifier

Chances are good that after a time you will tire of listening to your electronic creations on an earphone and will want to add an amplifier capable of operating a loudspeaker. That is our next project: an audio amplifier utilizing an integrated circuit (IC). Although the unit is simple and easy to build, it will provide loudspeaker output from any tuner that develops satisfactory volume on an earphone.

The amplifier uses an LM386 integrated circuit, which contains 10 transistors, 2 diodes, and 7 resistors, all in an 8-pin

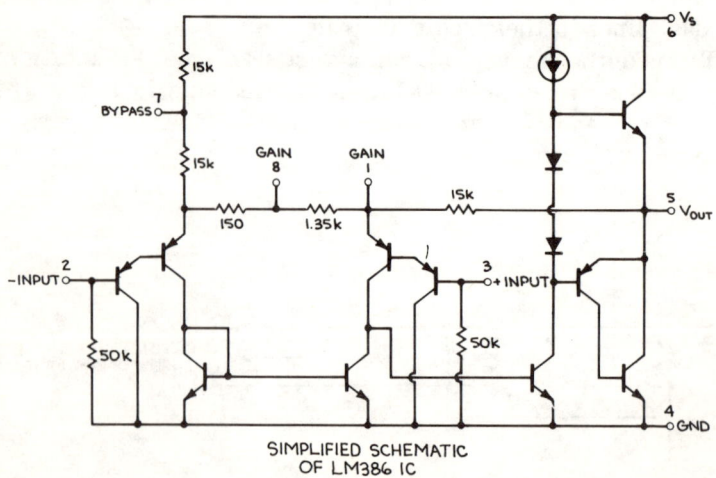

SIMPLIFIED SCHEMATIC
OF LM386 IC

plastic case. The IC is optimized for low current drain, which makes it a practical device for being powered by a 9-volt battery of the type used in small transistor radios.

## Shopping List

| Quantity | Description | Labeled in Diagrams |
|---|---|---|
| 1 | Type LM386 integrated circuit | |
| 1 | 10K potentiometer (volume control) with switch (Calectro B1-662 or equivalent) | R-1 |
| 1 | 470K, ¼ watt resistor | R-2 |
| 1 | 10 ohm, ¼ watt resistor | R-3 |
| 1 | 250 $\mu$F, 15 volt electrolytic capacitor | C-5 |
| 1 | 0.05 $\mu$F, 16 volt Mylar or ceramic capacitor | C-4 |
| 1 | 10 $\mu$F, 15 volt electrolytic capacitor | C-3 |
| 2 | 0.1 $\mu$F, 16 volt Mylar or ceramic capacitors | C-1, C-2 |
| 1 | Perforated board, 4½ × 5⅝ inches (11.4 × 14.3 cm) with 0.093-inch holes | |
| 2 | Cleats (see text) | |
| 4 | Screws for cleats | |
| 31 | Push-in terminals | |
| 2 | Fahnestock clips | |
| 2 | 6-32 machine screws (½ inch, 1.3 cm, long) with nuts | |
| 2 | Solder lugs | |
| 1 | 8-pin IC socket | |
| 1 | Battery connector (for 9 volt transistor battery) | |
| 1 | Knob for volume control | |
| 1 | 9 volt transistor battery (NEDA type 1604 or equivalent) | |
| | Hookup wire, No. 22 | |
| | Magnet wire, No. 26 (see text) | |

## PREPARING THE BASEBOARD

As in the earlier projects, the perforated board is mounted on wooden cleats, rubber feet, etc.—anything that will keep the board off the table. Here is the sequence for inserting the push-in terminals—the first step. All clips are inserted from the top of the board.

Fahnestock clips, fitted with soldering lugs, go in holes N-18 and R-18. The lugs are underneath the board.

| Row | Holes |
|-----|-------|
| B | 2, 7, 10, 13 |
| C | 1 |
| D | 6, 11, 13 |
| E | 3 |
| F | 8, 13 |
| G | 5, 6, 7, 8 |
| H | 13 |
| K | 3, 5, 6, 7, 8, 9 |
| M | 5, 6, 10 |
| N | 5, 8 |
| O | 6, 10 |
| V | 15, 19 |

The volume control goes at location I-17; enlarge the hole in the board enough to accept the threaded bushing on the con-

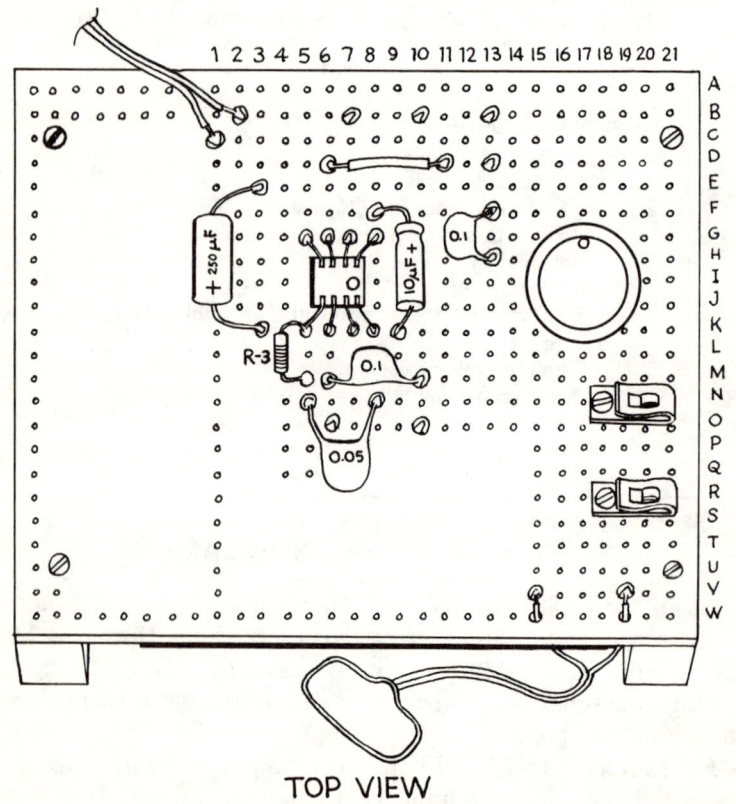

TOP VIEW

trol. Remove the nut and washer, insert the bushing through the hole, and replace the nut and washer. Be sure that the terminal lugs on the control are positioned as shown in the bottom-view drawing, and then tighten the nut (but not so tight that you strip the threads or crack the board).

### WIRING

The wiring is so simple and straightforward that you can complete it easily by following the pictorial drawings and referring to the schematic diagram. Note that the battery connection leads are looped through holes W-15 and W-19 and then soldered to clips V-15 and V-19 on top of the board.

Solder 470,000-ohm (470K) resistor R-2 between the solder lugs that are mounted to the Fahnestock clips.

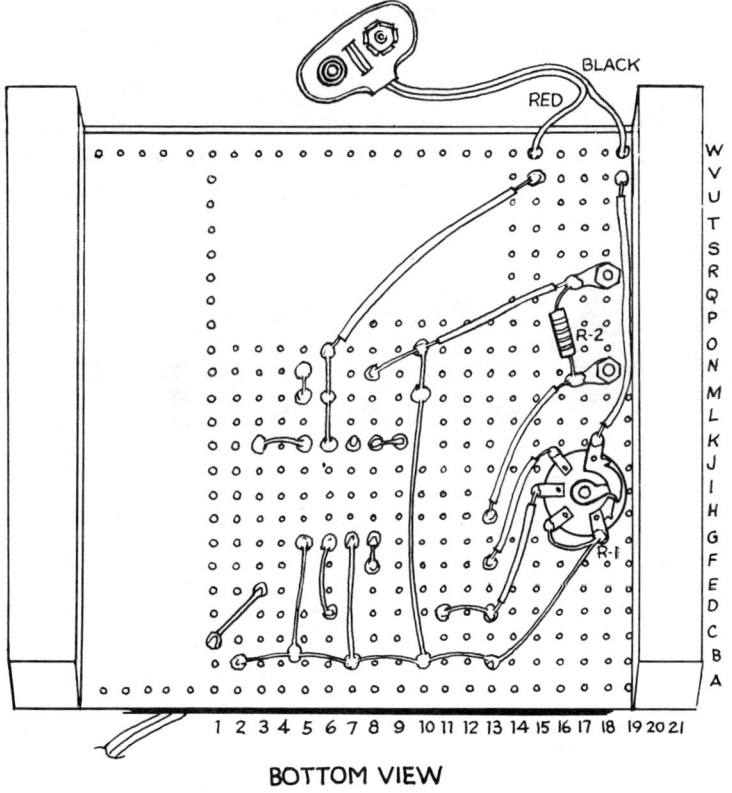

BOTTOM VIEW

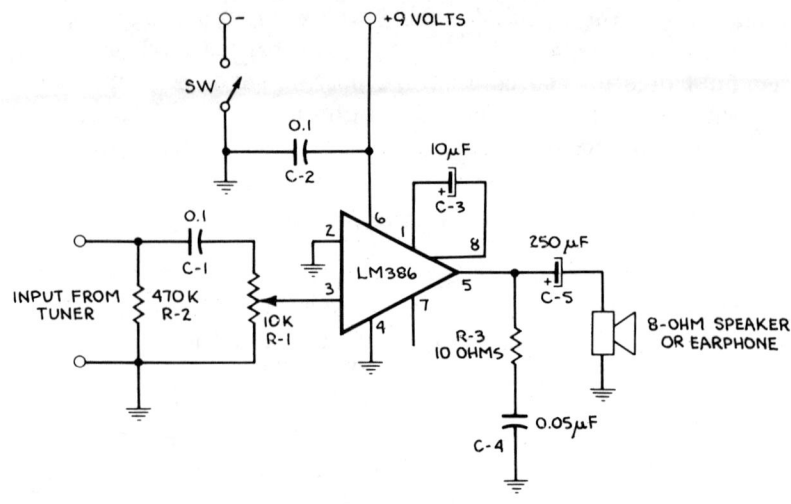

SCHEMATIC DIAGRAM

On top of the board, connect a lead between D-6 and D-11. A piece of 2-conductor speaker wire connects to terminals C-1 and B-2 for the loudspeaker connection. As shown in the photo, the lead can be tied down to the board with a couple of twists of bare hookup wire.

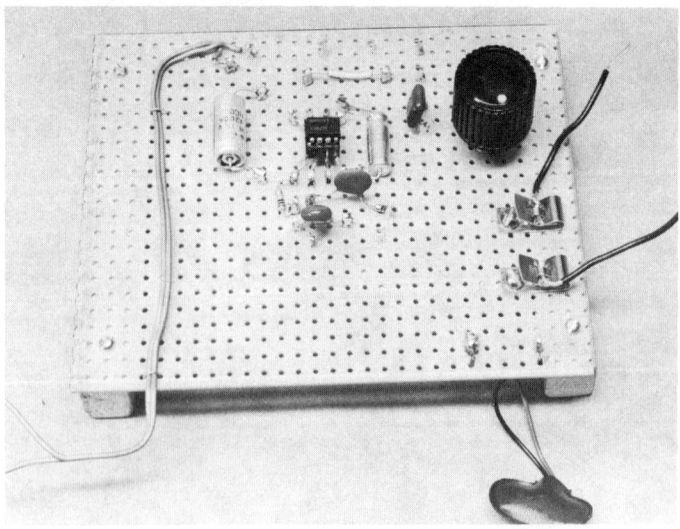

## ADDING PARTS

The only difficult task in building this little unit is that of soldering the 8-pin in-line socket to the push-in terminals. The leads on the socket are small and delicate, so the soldering job must be done carefully. However, there is a systematic way to go about it which will make the job somewhat easier.

1. Scrape (or better, sand with fine sandpaper) the enamel insulation from some magnet wire (for example, the No. 26 wire used for the Two Hour radio).
2. Cut some pieces about ¾ inch (1.9 cm) long, and with long-nose pliers or a soldering tool, form small loops just big enough to slip over the leads on the socket.
3. *Very* lightly, "tin" (add a thin coat of solder to) the loops.
4. Slide the loops over the socket leads, and bend the leads at right angles to the socket. Quickly, with a minimum of solder, solder the loop ends of the ¾-inch wires to the socket.
5. Finally, solder the opposite ends of the wires to the rows of clips.

The job will go better if you have tinned the clips before starting. The whole idea is to make the connections with a minimum of heat applied—enough to make the connection, but not so much that you unsolder the connections already made to the socket.

Examine your soldering job carefully. An "open" connection may result in the IC being destroyed when power is applied.

After you have accomplished the foregoing, the rest of the job is simple, indeed. Solder the remaining parts in place, again following the drawings.

## LOUDSPEAKER

This amplifier is intended for use with either an 8-ohm or a 16-ohm speaker. The drawings in this part of the chapter show the different speaker setups you can build yourself—one a "minimum" arrangement, the other far better but requiring some carpentry.

## Minimum Speaker

A small transistor-radio-type speaker can be mounted in a plastic box. The size of the box is not important, so long as it is large enough to accommodate the speaker (which should be at least 2 inches, and preferably 3 inches, in diameter). The drawing (page 61) shows how soldering lugs can be used to mount the speaker. Holes are drilled in the front to allow the sound to escape. The box shown has many small holes in it already. If you use a solid box, drill a row of ¼ inch (6 mm) holes in the back as well.

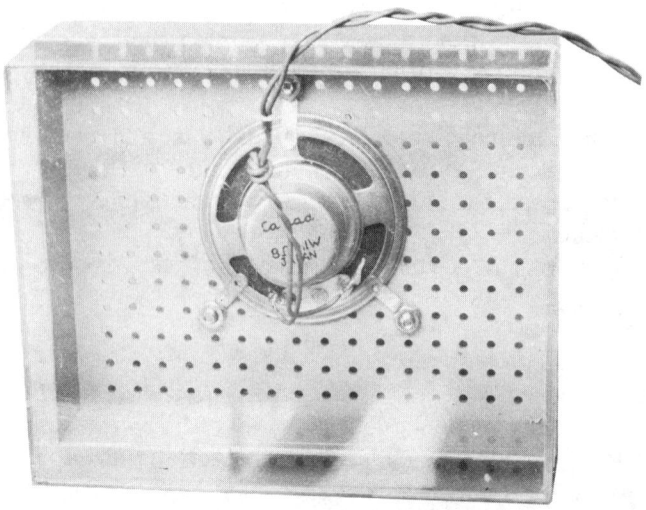

The quality from such a speaker will be equivalent to that of a small transistor radio. For *much* better quality, we can make up a speaker enclosure that will let us utilize a larger speaker.

## Low-Cost Speaker and Enclosure

A simple and low-cost way to provide a speaker for our amplifier is to utilize a 6 × 9 inch (21 × 23 cm) oval speaker of the type used in auto radio or stereo systems, and mount it in a homemade enclosure. The one shown in the photos and drawing can be placed on a shelf or table—or, better, be hung on the wall. Wall mounting will improve the tone.

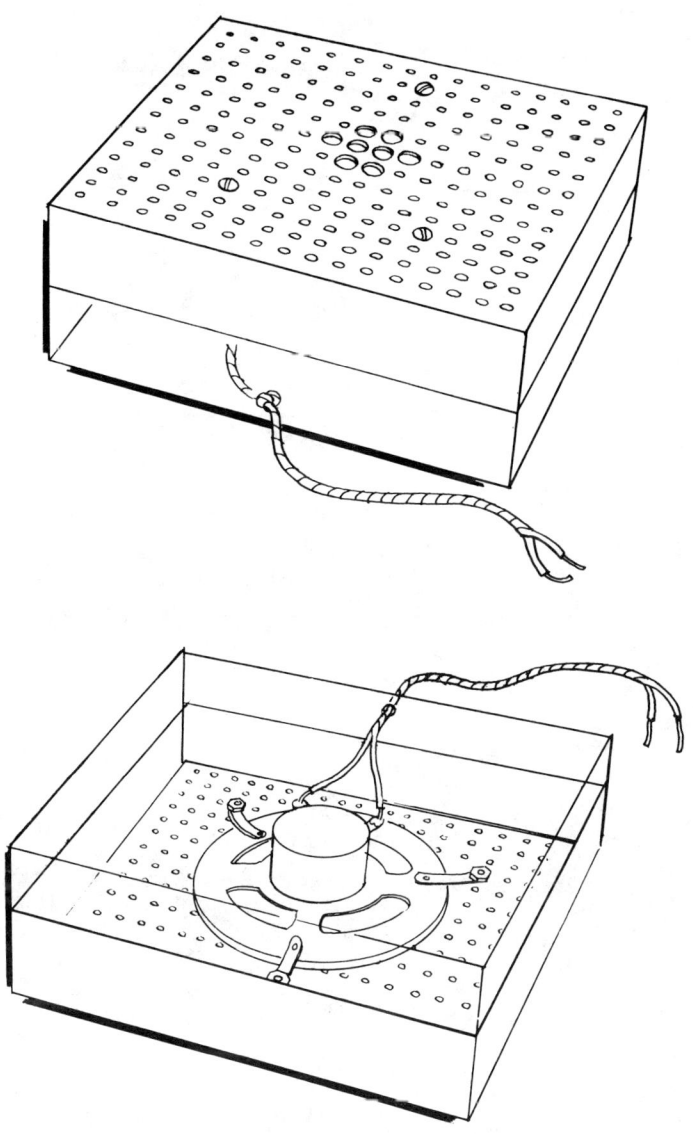

As shown in the drawing on page 62, the enclosure consists of a wooden frame and a tempered Masonite panel. The frame is made of standard "one-inch" Philippine mahogany, which actually measures approximately ¾ inch thick. If you do not have a power saw, you can do as the writer did, and have the

boards cut at a lumber yard. By getting good, squared cuts, assembly of the frame is much easier.

Put the frame together with finishing nails and white glue. The nail holes can be filled with a mahogany-colored filler ma-

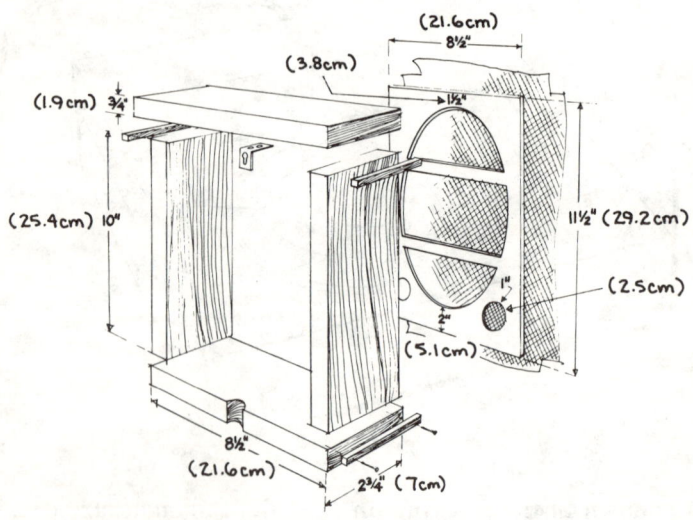

terial. Note that small "style strips" are added. These serve to cover the joint and act as "feet" if the enclosure is placed on a shelf.

The Masonite front panel is cut out to allow for sound passage. The *exact* size of the oval cut-out is unimportant; just make it somewhat smaller than the speaker cone. The dimensions in the drawing will be okay for most speakers.

After the cutouts are made, cover the panel with speaker grille cloth. Actually, any loosely woven cloth could be used. As shown, the cloth is wrapped around the edge of the board and glued to the panel.

The mahogany frame (pine will do, mahogany is just more attractive!) should be stained, painted, etc., before the front panel is added. The panel can be affixed to the frame with glue and very small finishing nails. If you use small nails, they will tend to bury themselves in the grille cloth and not be visible from the front. If you want to give the enclosure an even more professional look, add a "frame" of wood molding on the front panel.

A small metal angle bracket is mounted on one end of the enclosure so that it can be hung on the wall or on a picture hook. Note that there is a half-round slot on the frame (put there with a round file) so that the speaker extension wire can

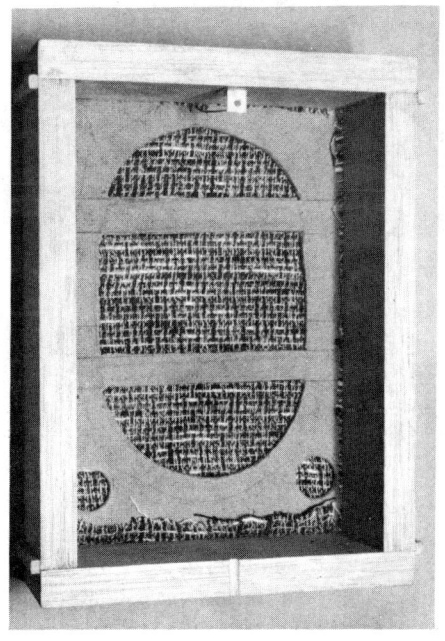

be run down the wall without pushing the speaker away from the wall.

The 8-ohm speaker itself is mounted to the front panel by means of 6-32 ornamental machine screws and nuts. Use washers or grommets under the nuts, since most speakers have quite large mounting holes. You can use the speaker itself for determining where the holes should be—however, be careful not to punch a hole in the speaker cone!

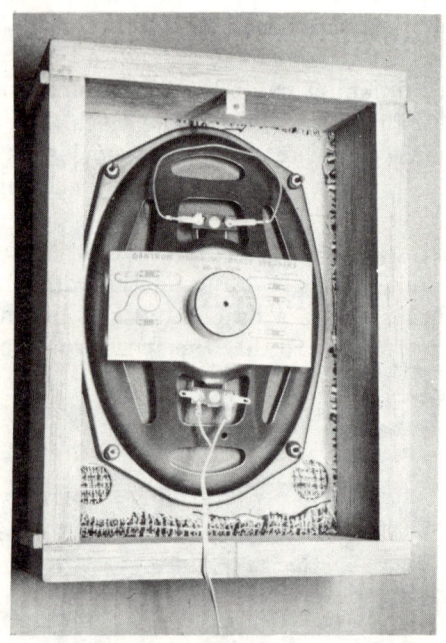

## A WORD ABOUT SELECTING SPEAKERS

Most 6 × 9 inch replacement speakers are similar in design, usually 8 ohms. There is one type, like the one in the photos, which can be adjusted for several impedances—for example, 5, 10, 20, and 40 ohms. Such a speaker is fine; just follow the instructions that come with the speaker for making the connections for 10 ohms, which is close enough.

You may find that different speakers have different magnet weights. As a rule of thumb, the heavier the magnet, the better the speaker.

## CHECKING THE AMPLIFIER

We will assume that you want to use the amplifier with one of the two tuners described in previous chapters, and hence you have a ready source of signal for the amplifier input. If not, see Chapter 12 for a way of testing the amplifier with a screwdriver as a "signal source."

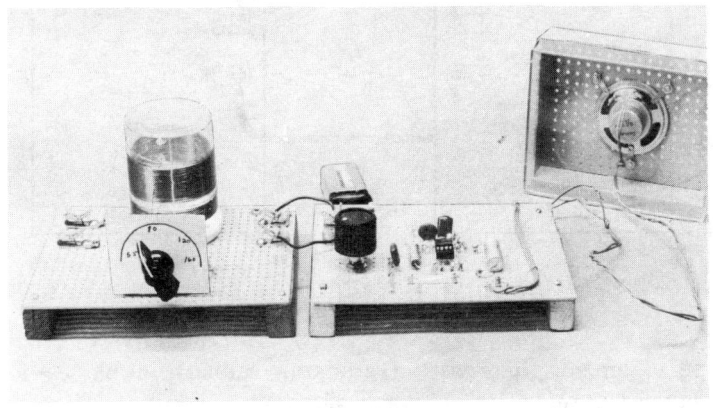

The photo and block diagram show how to hook the two units together—simplicity itself, since it involves connecting opposite pairs of Fahnestock clips. Resistor R-2 in the amplifier takes the place of the earphone in providing a "load" for the diode detector.

Plug the IC into the socket. Be careful; the pins on the IC are small and fragile. Also—and *VERY IMPORTANT*—note that the IC *could* be inserted either of two ways; the 8 pins are identical. The *KEY* is a small circle on top of the plastic case of the IC. This circle tells you which end is which. Insert the IC in the socket so that the circle is *TOWARD* the volume control. A mistake at this point could ruin the IC.

Now hook up the speaker to the speaker leads. Be certain you make good connections—it is important that the speaker be in place to provide a "load" for the IC.

Connect the 9-volt transistor battery. Turn the volume control knob so that the switch clicks on. As you advance this control, the volume should increase, providing you have the Two Hour radio or other signal source tuned to a station.

The amplifier has no effect on the tuning. It simply makes the receiver signal louder. Tune the set exactly as you did when you had it hooked to an earphone. Speaking of earphones, with this unit you *can* use an 8-ohm earphone (of the

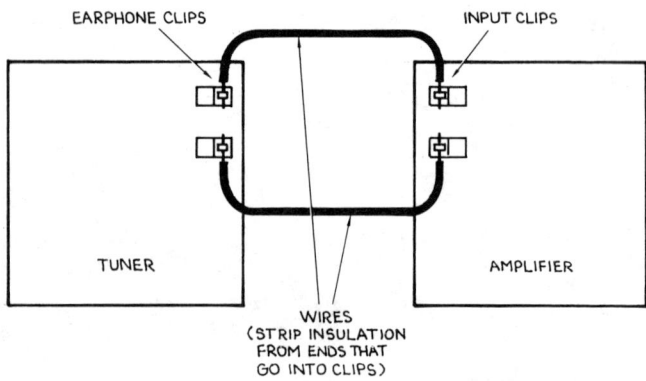

type often supplied with transistor radios) in place of the speaker for reception of weak signals.

## TROUBLESHOOTING

The circuit uses so few parts that should it fail to operate, the odds are very strong that the trouble is the result of a poor soldering job or a wiring mistake. The chance of having a bad part is very slim. The IC is the most likely candidate, but a faulty one is not probable.

Because soldering to the socket is a delicate and tricky job, that would be the first place to look for trouble. Examine the connections carefully, ideally under a magnifying glass.

Be sure that signal is coming *into* the amplifier. To check this, touch the leads from your crystal earphone to the earphone clips. You should hear a signal almost as strong as if the tuner were not connected to the amplifier.

Tested in the writer's workshop, the amplifier would overload the "minimum speaker" with the volume control advanced all the way. It would drive the wall-enclosure speaker to room-filling volume. In both cases, the Two Hour radio was the signal source.

**CHAPTER 6**

# Transistor AM Receiver

In previous chapters, we covered the building of two very simple tuners. Both of them depend entirely upon the signal that is pulled in by the antenna and ground system; they simply transform this signal into an audio signal which we can hear. No power of any kind is utilized to *increase* the incoming signal from the radio station; we simply rectify the signal with a diode. This means that at least a passable antenna is necessary for us to hear the local station or stations.

By utilizing an *audio* amplifier like that described in Chapter 5, the incoming signal can be increased by amplifying it *after* it passes through the diode detector. Another approach is to step up the signal *before* it reaches the diode detector, thus greatly increasing the *sensitivity* of the set. Such an amplifier is called a *radio frequency* (rf) amplifier. Chapter 7 describes a set which utilizes this method.

In this chapter, we will cover a unit which is something of a compromise. It uses a regenerative detector circuit, which in practical terms provides greatly increased rf sensitivity, detection, and audio amplification all in one stage. It utilizes a transistor and is powered by a battery. The big advantage of this unit is that it will operate on a *small* antenna, an 8- or 10-foot piece of wire being adequate to pull in local stations. It also has quite a bit of inherent selectivity (the ability to tune in just *one* station at a time).

As always in the case of electronic gear, this set is something of a tradeoff. While the unit is *far* more sensitive than the two previously described sets, it is more complicated to build, and it is not capable of as "high fidelity" reception as is the "Selective AM Tuner" in Chapter 4. However, it is an ideal "next step" because it will give you some practice in building that will enable you to tackle more complicated projects in subsequent chapters. And—as mentioned previously—it is the *best* bet (other than the unit described in Chapter 7) for you

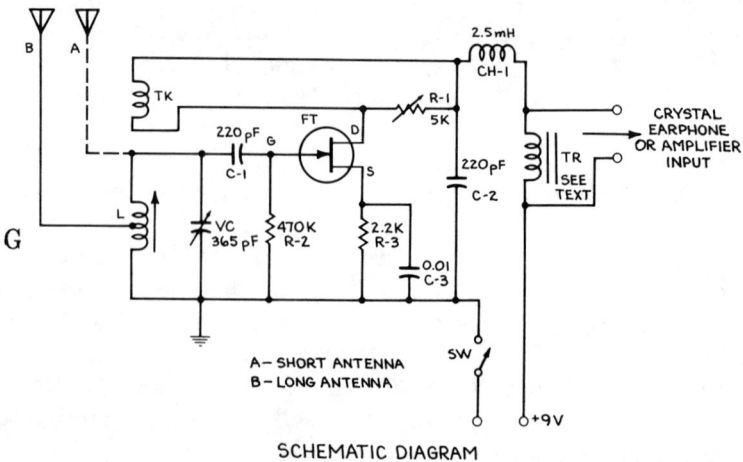

SCHEMATIC DIAGRAM

if you do not have space for some sort of outside antenna system.

## Shopping List

| Quantity | Description | Labeled in Diagrams |
|---|---|---|
| 1 | Perforated board, 4½ × 5⅝ inches (11.4 × 14.3 cm) with 0.093-inch holes | |
| 20 | Push-in terminals for 0.093-inch holes | |
| 8 | Solder lugs | |
| 5 | Fahnestock clips | |
| 10 | 6-32 machine screws, ½ inch (1.3 cm) long, with nuts | |
| 2 | Cleats (see text) | |
| 4 | Screws or bolts for cleats | |
| 1 | Variable capacitor, 365 pF with dial (Calectro A1-232 or equivalent) | VC |
| 1 | 5000 ohm miniature volume control with switch (Calectro B1-661 or equivalent) | R-1 |
| 1 | Radio frequency choke, 2.5 mH | CH-1 |
| 1 | Miniature or subminiature audio transformer, 100,000-200,000 ohm primary preferred, but 10,000 ohms acceptable (Calectro D1-710 or equivalent is suitable.) | TR |
| 1 | Ferrite-core antenna (Calectro D1-841 or equivalent) | L |
| 2 | 220 pF mica capacitors | C-1, C-2 |
| 1 | 0.01 μF, 16 volt Mylar or ceramic capacitor | C-3 |
| 1 | 470K, ¼ watt resistor | R-2 |
| 1 | 2.2K, ½ watt resistor | R-3 |
| 1 | Field-effect transistor, type MPF102 or equivalent | FT |
| 1 | Battery, 9 volt (NEDA type 1604 or equivalent) | |
| 1 | Battery connector | |
| 1 | Battery holder | |
|  | Hookup wire | |

## CONSTRUCTION

We will start building by inserting the push-in terminals in the board and also mounting the various soldering lugs and

Fahnestock clips. This procedure will provide guideposts for our wiring and help avoid wiring errors.

As in Chapter 4, the holes in the board are identified by letter and by number so that you can locate them easily. As mentioned previously, it is a good idea to write the letters and numbers directly on the board opposite the rows of holes.

The clips and terminals will be easier to mount if some kind of cleat or bracket is provided so that the board clears the table upon which it rests. The volume control (actually a variable resistor for controlling regeneration) requires a bit of space below the panel, so be certain that the mounting cleats are high enough.

**Push-In Terminals**

Terminals are pushed into the board from the bottom at F-8 and I-8 (or to fit the transformer). The others are pushed in from the top as follows:

| Row | Holes |
|-----|-------|
| E | 10, 13, 17 |
| G | 13, 15 |
| J | 10, 16 |
| K | 11, 13, 15, 17 |
| L | 11, 14 |
| M | 13 |
| N | 10, 12, 14, 17 |

**Soldering Lug and Clip Terminals**

Enlarge holes C-4, C-8, C-12, C-17, C-20, and H-19 with a $5/32$-inch drill bit. Then mount:

Clip FH-0 with soldering lug-T-5 underneath board
Clip FH-1 with soldering lug T-4 underneath board
Clip FH-2 with soldering lug T-3 underneath board
Clip FH-3 with soldering lug T-2 underneath board and lug T-8 on top of board
Clip FH-4 with soldering lug T-1 underneath board

In hole H-19, mount lug T-7 on top of the board and T-6 underneath the board with one bolt.

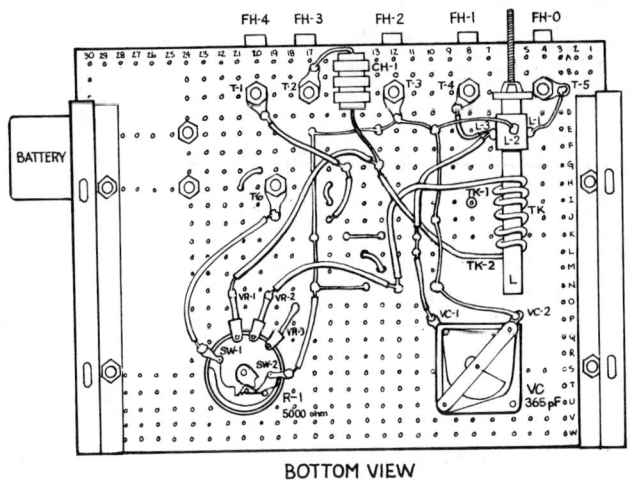
BOTTOM VIEW

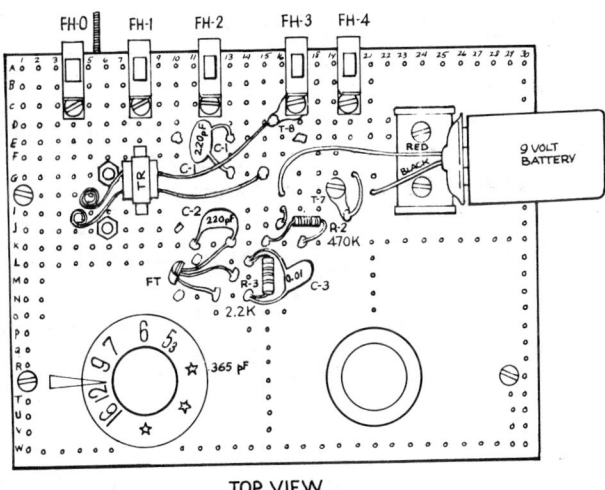
TOP VIEW

## Wiring Underside of Board

On the underside of the board, use hookup wire to connect the terminals as follows. (S means solder; DS means don't solder because another wire will be connected here in another step.)

N-14 (S) to N-17 (DS)
N-17 (DS) to K-17 (DS)

K-17 (S) to E-17 (DS)
E-17 (S) to E-13 (DS)
E-13 (S) to lug T-3 (DS)
lug T-3 (S) to E-10 (DS)
E-10 (DS) to J-10 (DS)
J-10 (S) to N-10 (DS)
G-15 (DS) to J-16 (S)
K-15 (S) to K-13 (S)
M-13 (S) to L-14 (S)
lug T-1 (S) to G-15 (S)

**Mounting of Parts**

Mount the coil bracket. This is usually packaged as a small metal strip which can be bent into an "L" shape. The short part of the L contains the hole into which the coil snaps. The long part of the bracket is bolted to the baseboard through enlarged holes G-6 and J-6.

Cut mounting holes for the 365-pF variable capacitor (enlarge hole S-8) and the 5000-ohm variable resistor (volume control) (enlarge hole S-21).

Mount the variable capacitor and the variable resistor.

Mount the battery holder, using 6-32 bolts.

Snap the coil into the mounting bracket after rotating it so that the number 2 terminal (center terminal when the coil is viewed from the end) is in a vertical position. Turn the board over.

Mount the small transformer (TR) by pushing it down on clips F-8 and I-8, and then soldering it to the clips. (Very small bolts and nuts could be used instead of clips and soldering.)

**Additional Wiring**

Now, we'll start wiring again. Follow the chart below for installing the small parts on top of the board. Solder carefully. Be certain you have a connection; sometimes solder will run down into the clips *without* making a connection to the part.

| Part | Connect to |
|------|------------|
| C-1 | G-13 (S) and E-13 (S) |
| C-2 | K-11 (S) and K-13 (DS) |
| C-3 | L-14 (DS) and N-14 (DS) |
| R-2 | K-15 (S) and K-17 (S) |

| | |
|---|---|
| R-3 | L-14 (S) and N-14 (S) |
| Battery Connector Leads | Thread red wire down through hole H-16, back up through hole I-16, and solder to J-16. Thread black wire down through hole H-21, back up through hole G-20, and solder to solder lug T-7. |

As shown in the bottom-view drawing, there is a 6-turn "tickler" winding added around the tuning coil. Form this coil out of hookup wire over a round surface (for example, the handle of the soldering aid tool) so that you can slip it easily over the coil. *Important:* Note that the coil (TK) is wound *counterclockwise*. With most tuning coils, this will ensure the desired regeneration—but *not* always. More on that later.

Now continue wiring on the bottom of the board.

| *Part* | *Connect* |
|---|---|
| TK | TK-1 to N-12 (DS) |
| | TK-2 to G-13 (DS) |
| CH-1 | To solder lug T-2 (S) and G-13 (DS) |
| R-1 | VR-1 (S) to G-13 (S) |
| | VR-2 (S) to N-12 (S) |
| SW | SW-1 (S) to Lug T-6 (S) |
| | SW-2 (S) to N-17 (S) |
| L | L-1 (S) to Lug T-5 (S) |
| | L-2 (S) to E-10 (S) |
| | L-3 (DS) to Lug T-4 (S) |
| | L-3 (S) to K-11 (DS) |
| | K-11 (S) to L-11 (DS) |
| VC | VC-1 (S) to L-11 (S) |
| | VC-2 (S) to N-10 (S) |
| | (When soldering to the terminals on the variable capacitor, follow the precautions outlined in the previous chapter.) |

The final above-chassis part we will connect is the small transformer, TR. Note that only *one* pair of leads (one winding) is used. Normally, these transformers come with a chart that indicates which winding is which. We want to use the *high-impedance* winding—10,000 ohms impedance or more. The color coding is not important—simply select the high-impedance winding. Solder one of the leads from this winding to

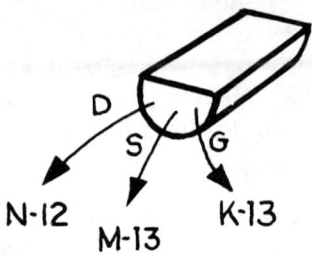

solder lug T-8 and the other to the push-in terminal at G-15. Coil up the unused leads so that they are out of the way but still available in case you want to use the transformer later in another project. (Be sure the ends of the leads don't short together.)

Now be careful. The next step is to mount the transistor, FT. Like the diodes utilized in the previous sets, a transistor is a delicate device and needs protection from heat. But first, we must identify the leads—"G," "D," and "S." These are illustrated in the drawing.

"Tin" the leads by applying solder on the ends one at a time. Keep your long-nose pliers on each lead *between* the transistor body and the tip of the soldering iron, just as was done with diodes in previous chapters. Once the leads are shiny with solder, you are ready to connect the transistor to the proper terminals.

Again, protect the transistor with your pliers or, easier, with a "heat sink" tool like that illustrated in Chapter 1. This tool is a kind of spring-loaded clip that will grip the lead and absorb heat while the soldering is going on. Solder the transistor leads as follows:

>G to clip K-13
>D to clip N-12
>S to clip M-13

## TESTING THE UNIT

Now check—and recheck—the wiring. Examine *all* soldered connections, especially push-in terminals that have been soldered from both above and below the board. If possible, examine all connections with a small magnifying glass.

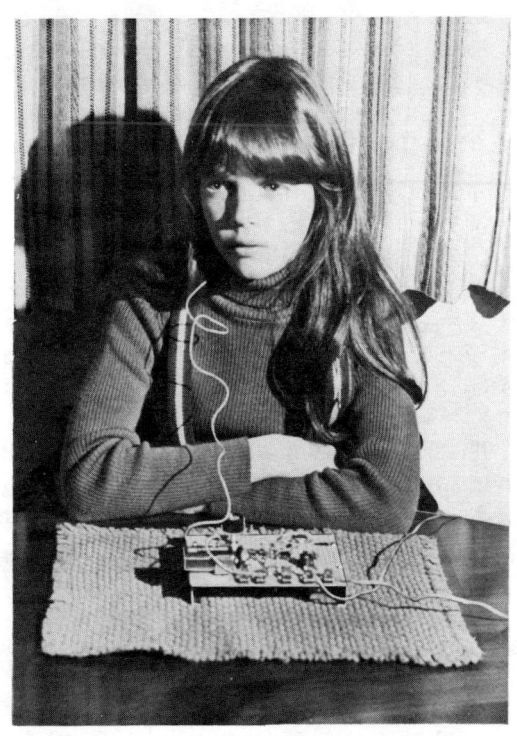

As a final check, lay a piece of tracing paper over the pictorial drawings, and with a red pencil draw in each lead as you examine it. The objective is to make certain that you have omitted nothing and that every wire goes where it *should* go.

Satisfied? Hook up the crystal earphone to FH-3 and FH-4. Connect whatever "ground" is available to FH-2. If you have an outside antenna, hook it to FH-0. Otherwise, connect a 6- to 10-foot piece of wire (hookup wire is okay) to FH-1.

Snap the battery connecter onto the battery. Rotate the variable resistor to about midrange.

Now turn the tuning capacitor. You *should* hear a high-pitched whistle each time you tune past a station. Tune in a strong whistle, then back off the variable resistor until the whistle stops. Readjust the variable tuning capacitor for best reception. The objective is to tune the station right on the nose, and use as much "regeneration" as possible *short* of creating the whistle that occurs as the set goes into oscillation.

75

Properly controlled, regeneration increases sensitivity tremendously.

## NO WHISTLES?

This could mean a wiring mistake or a bad soldered connection. Most probably, it means that the "polarity" of TK is wrong. Unfortunately, there is no way of anticipating this because different manufacturers of coil "L" make their coils differently. But the solution is easy. Disconnect TK-1 and TK-2. Switch them. For example, if TK-1 formerly went to N-12, connect it to G-13—and connect TK-2 to N-12. This reverses the polarity.

Try again. The odds are that—assuming you have been careful in the wiring job—the set will perform as it should.

Yes, the little tuner will power the IC amplifier from Chapter 5. Simply hook the headphone terminals to the input terminals of the amplifier. Just be certain that terminal FH-2 on the tuner goes to the terminal at R-18 on the amplifier. Terminal FH-3 on the tuner should connect to the terminal at point N-18 on the amplifier.

Keep the *volume* down. Otherwise, if you have the tuner oscillating, the whistle may bring complaints from the other end of the house!

CHAPTER 7

# Hi-Fi AM Tuner

The Selective AM Tuner described previously has the shortcoming of requiring an outside antenna. The Transistor AM Receiver will operate on a very modest indoor antenna, but it requires some skill to operate, and it leaves a bit to be desired in terms of fidelity. The unit to be described next overcomes both of these problems. It has a self-contained antenna, and it is capable of good fidelity. Further, it has a characteristic called agc, or "automatic gain control," a highly useful technique for holding the volume at a nearly constant level.

The set utilizes a linear integrated circuit (IC) that looks for all the world like a transistor. Actually, it is far more so-

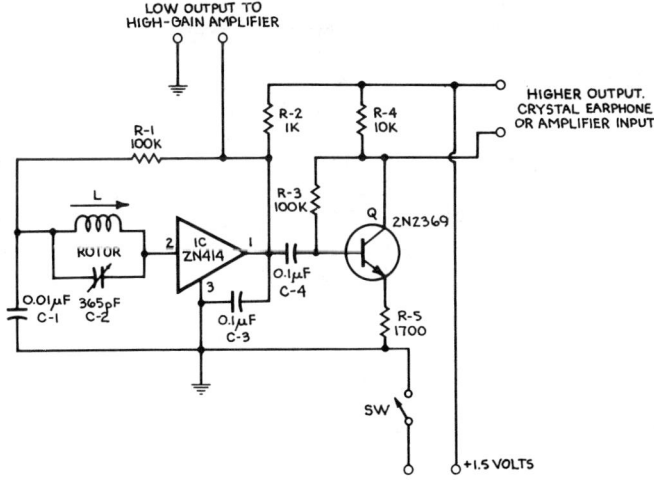

77

phisticated than that; in fact, there are 10 transistors in a tiny package that provides a highly sensitive, multiple-stage rf amplifier, detector, and agc. The integrated circuit, ZN414, has been called a "complete radio in a chip." In practical

## Shopping List

| Quantity | Description | Labeled in Diagrams |
|---|---|---|
| 1 | Linear integrated circuit, radio receiver (Ferranti type ZN414) | IC |
| 1 | Transistor, npn (2N2369 or equivalent) | Q |
| 1 | Ferrite core antenna coil (Calectro type DL841 or equivalent) | L |
| 1 | Single-pole, single-throw (spst) toggle switch | SW |
| 1 | 1700 ohm, ¼ watt resistor (color code brown, violet, red) | R-5 |
| 1 | 10,000 ohm (10K), ¼ watt resistor (color code brown, black, orange) | R-4 |
| 2 | 100K, ¼ watt resistor (color code brown, black, yellow) | R-1, R-3 |
| 1 | 1000 ohm (1K), ¼ watt resistor (color code brown, black, red) | R-2 |
| 1 | 0.01 µF, 15 volt (or more) Mylar or ceramic capacitor | C-1 |
| 2 | 0.1 µF, 15 volt (or more) Mylar or ceramic capacitor | C-3, C-4 |
| 1 | 365 pF variable capacitor with dial (Calectro A1-233 or equivalent) | C-2 |
| 1 | Perforated board, 2 × 5⅝ inches (5.1 × 14.3 cm) with 0.093-inch holes | |
| 7 | Soldering lugs | T |
| 6 | Push-in terminals for 0.093-inch holes | |
| 14* | 6-32 machine screws, ½ inch (1.3 cm) long, with nuts | |
| 2 | Terminal strips, 2-terminal, screw type | |
| 1 | Battery holder for single AA cell | |
| 2 | ½ × ½ inch angle brackets | |
| 1 | Plastic box with aluminum lid, 7⅞ × 4½ × 2½ inches (20 × 11.4 × 6.4 cm) | |
| | Two-conductor speaker wire (approx. 2 feet) | |
| | Hookup wire, No. 22 | |

*Number required may vary, depending on needs for angle brackets, battery holder, coil bracket, etc.

terms, this IC makes possible a very simple tuner that tunes like an ordinary radio (it requires no critical regeneration control) and has the important plus advantage of being capable of high fidelity—*far* better than most ordinary radios, and better than the am section of at least some am-fm tuners.

Actually, designers don't worry too much about fidelity in the am portion of the typical am-fm tuner, because the assumption is made that the user will be interested primarily in fm stereo, the am portion being provided only as a convenience for listening to the news, etc. In a way, this is a shame, for while the typical am broadcasting station does not match an fm station in fidelity, it is capable of much better sound than most people ever hear from an am radio. If you build this tuner, and then use it to drive a good quality amplifier, the quality is almost certain to surprise you.

## WIRING

In building most of the units in this book, the parts are mounted above a board with the wiring underneath. There is nothing sacred about this approach; in fact, some sophisticated printed-circuit-board layouts have wiring and parts on both top and bottom.

In our case, we will start by inserting the push-in terminals and mounting the soldering-lug terminals from the underside of the board. Here is the sequence to follow:

| Row | Hole |
|---|---|
| | Terminal Lugs |
| B | 10 (Bolt also secures battery holder to top of board), 16 |
| D | 3, 15, 24 |
| I | 29 (Same bolt holds two lugs, one on bottom of board, one on top.) |
| | Push-In Clips |
| G | 12, 15 |
| H | 11, 16 |
| I | 12, 15 |

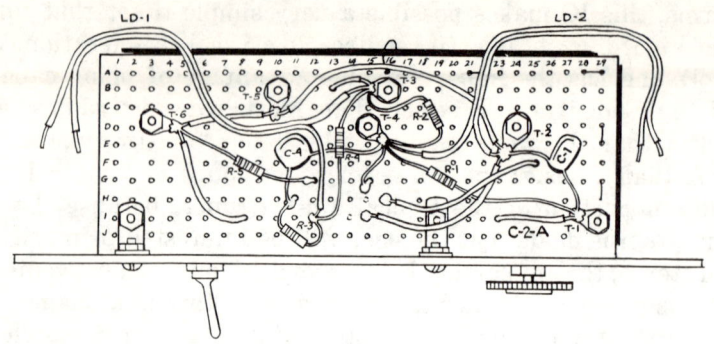

With the clips and lugs in place, we can start wiring.

- [ ] T-6 (DS) to T-5 (DS)
- [ ] T-5 (S) to T-2 (DS)
- [ ] T-2 (DS) to I-15 (S)
- [ ] T-4 (DS) to G-15 (S)

Connect

- [ ] R-3 between H-11 (DS) and I-12 (DS)
- [ ] C-4 between H-11 (S) and T-4 (DS)
- [ ] R-5 between T-6 (DS) and G-12 (S)
- [ ] R-4 between I-12 (DS) and T-3 (DS)
- [ ] R-1 between T-1 (DS) and T-4 (DS)
- [ ] R-2 between T-3 (DS) and T-4 (DS)
- [ ] C-1 between T-1 (DS) and T-2 (DS)

Install wire leads as follows:

- [ ] Connect a 4-inch lead to terminal T-6 (S), and then poke the wire through hole I-8.
- [ ] Connect a 4-inch lead to terminal T-1 (S), and then poke the wire through hole I-23.
- [ ] Connect a 2-inch lead to terminal T-3 (DS), and then poke the wire through hole B-13.
- [ ] Connect a 4-inch lead to H-16 (S), and then poke the wire through hole F-25.

Connect the two-conductor speaker wire:

- [ ] Silver end of LD-1 to T-3 (S).
- [ ] Copper end of LD-1 to I-12 (S).

☐ Silver end of LD-2 to T-2 (S).
☐ Copper end of LD-2 to T-4 (S).

Be sure that none of the bare lead wires from the parts touch any other bare lead wires or metal parts.

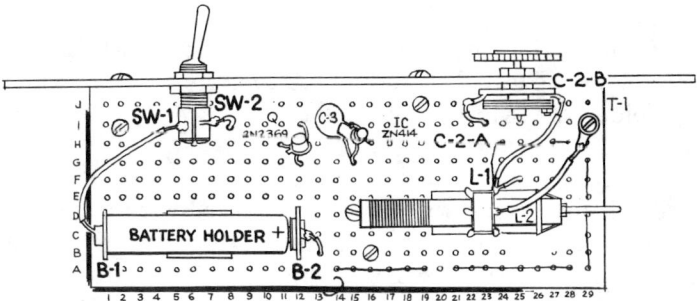

There are several parts mounted on top of the board. These include the battery holder and a lug for T-1 as mentioned above, and coil L. The coil is mounted as were similar coils used in Chapter 4.

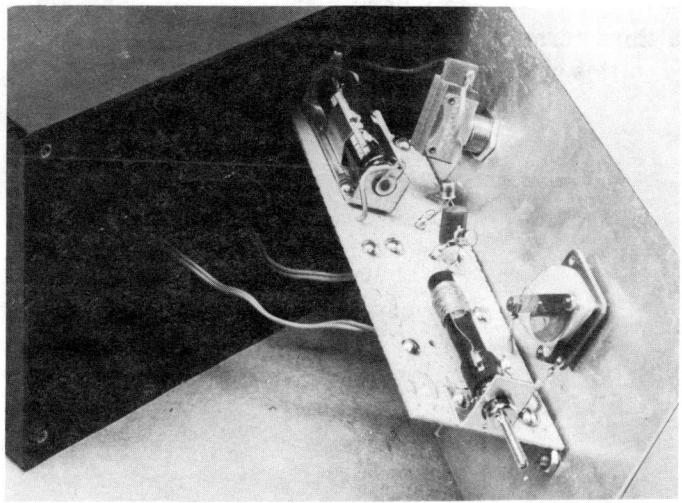

The variable (tuning) capacitor and the switch are mounted on the front panel. The board also is mounted to this panel by means of the small metal angle brackets. Use the photos and drawings as a guide for locating and drilling the mounting holes for installing the particular parts you have.

Now you can wire the remaining parts on top of the board.

- [ ] Solder a lead between soldering lug T-1 and L-2.
- [ ] Solder a 1½-inch lead to C-2-B (protect capacitor from heat).
- [ ] Connect the other end of the same lead to L-1 (DS).
- [ ] Solder the lead from hole F-25 to L-1.
- [ ] Solder the lead from hole I-23 to C-2-A (protect capacitor from heat).

The coil may come with a short piece of wire connected to terminal L-1. This piece of wire can be threaded in and out of the holes in the board (see drawings), but this wire must not touch any parts.

Most coils of this type have the same terminal connections, but, unfortunately, some do not. However, there is usually some kind of diagram with the coil that will help you make the proper connections. What is vital is that the "ground" end of the coil goes to T-1. The other end, sometimes labeled "grid" goes to C-2-B, the stationary plate of the variable capacitor. The third terminal, sometimes called "transistor tap," is not used in this circuit.

2N2369
BASE H-11
I-12 C
G-12 E

IC ZN414
H-16 INPUT
OUTPUT 1 2 3 I-15
GROUND
G-15

The connections to the switch and battery holder are shown in the drawing.

We are almost finished. Solder the 0.1-$\mu$F Mylar capacitor (C-3) between G-15 and I-15.

Now be careful. Using the techniques described in Chapter 6, protect the leads on the ZN414 with a heat sink. Using the diagram, identify the leads—and *don't* make a mistake. Carefully solder the "input" (lead 2) to H-16. In similar fashion, solder the "output" (lead 1) to G-15. "Earth" (ground) (lead 3) solders to I-15.

Wire in the transistor. Again, check the lead connections in the drawing. "E" goes to G-12. "Base" solders to H-11. "C" goes to I-12.

### TESTING

The writer always prefers, if possible, to test equipment before mounting it in a cabinet. This is easy to do with a unit like the Hi-Fi Tuner chassis, which will function simply resting on a table.

1. "Tack" solder the leads from a crystal earphone to LD-1.
2. Plug a penlight cell into the battery holder. Observe polarity—notice that the "plus" end is farthest from the end of the board.
3. Snap the switch on, and turn the tuning capacitor. You have a 50-50 chance of picking up a station (depends on whether the switch was actually on or off). Hear nothing? Try the switch the other way. This time—barring a wiring mistake—you should pick up a station as you rotate the dial.
4. Assuming there is a station at approximately 550 kHz, a station should come in at about that frequency on the dial (tuning capacitor almost closed). If not, tune in a station that has a known frequency near the low end of the dial, turn the dial to the frequency of the station, and then adjust the slug with a screwdriver (see Chapter 4) so that the station comes in again. All of this is an approximation because it will depend on how you affix a pointer to the front panel.

You will notice that the set is somewhat directional—turning it in a different direction will increase or decrease the signal. This phenomenon is characteristic of almost any set that has an antenna coil with a ferrite core.

## ASSEMBLY

If the set does not work, recheck the wiring wire by wire and part by part. One good technique for checking is to lay a semitransparent piece of paper over the drawing and then, with a colored pencil, draw in each wire and part as you find it. Be particularly thorough in checking connections to the transistor and integrated circuit.

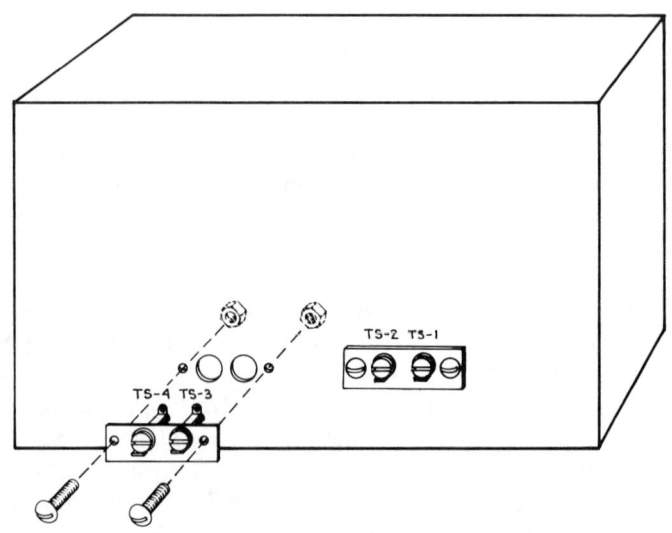

Set working fine? Now prepare the cabinet. Mount two terminal strips in the rear of the cabinet for connection to LD-1 and LD-2. You will need to make holes in the rear of the box for the mounting bolts and to clear the terminal lugs and screws. Be sure the location you choose places the terminal strips where they will not interfere with the circuit board when it is mounted in place.

Solder LD-1 (silver lead) to TS-1
Solder LD-1 (copper lead) to TS-2
Solder LD-2 (silver lead) to TS-3
Solder LD-2 (copper lead) to TS-4

Now, affix the panel to the cabinet with screws as shown in the drawings.

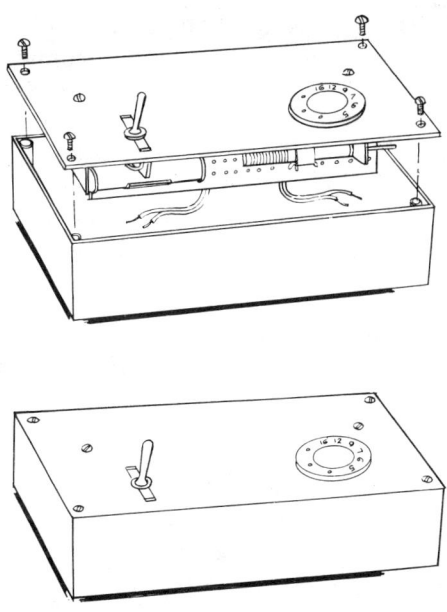

A crystal earphone can be connected to TS-1 and TS-2. Also, the tuner can be used to drive the IC amplifier described in Chapter 5 to provide sufficient output to operate a speaker with good volume.

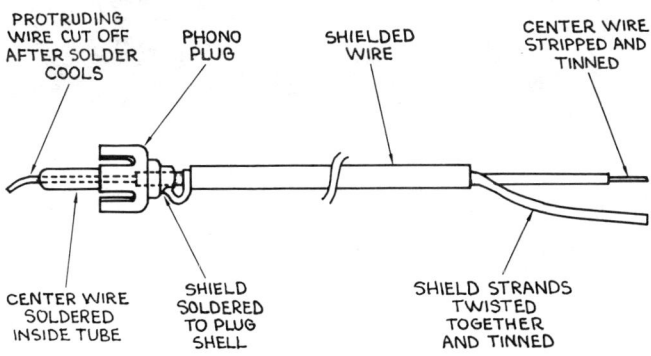

For using the tuner with a stereo receiver-amplifier or other hi-fi amplifier that has provision for "external input," you need to prepare an adapter similar to the one above. For most sets you will use the TS-1, TS-2 connection (shield connected to TS-1). However, if the amplifier with which you are working has a lot of gain, you may be able to connect to TS-3 and TS-4 (shield connected to TS-3), which provides a signal directly from the IC chip, eliminating the very small distortion introduced by the transistor amplifier stage. Unless you really have a good ear, chances are good you won't be able to tell the difference. Either way, the quality should be excellent. The photo shows the tuner connected to the utility amplifier described in Chapter 10.

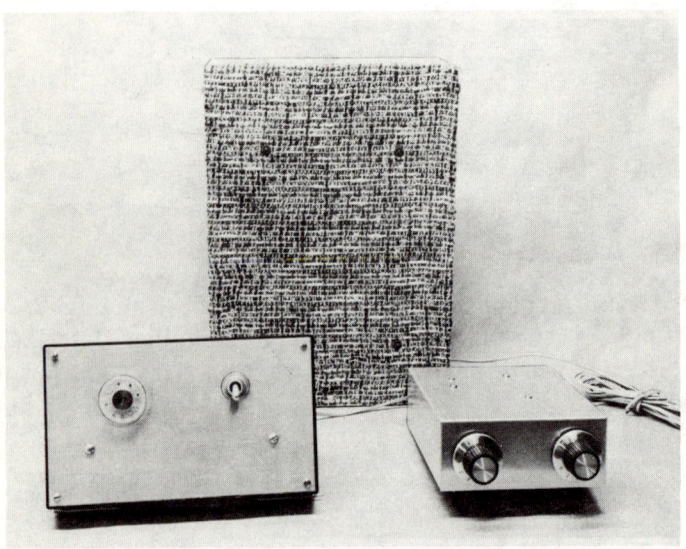

**CHAPTER 8**

# About Parts and Symbols

When you are starting out in electronics, pictorial diagrams like those scattered throughout this book are a big help. They make it easy for you to duplicate the original set with a minimum of technical knowledge. But in so doing, they limit you to following specific instructions for a specific set. Later, you will be unable to tackle the many interesting building projects that appear in magazines with circuit (schematic), but no pictorial, diagrams. Likewise, you will never be able to do any serious electronics work—for example, radio or tv servicing—or get a general-class radio amateur license. These activities require you to be able to read circuit diagrams.

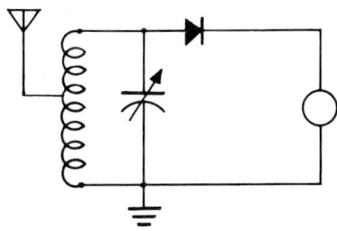

Actually, most old hands in electronics groan mightily when they have to check through a pictorial diagram. They have found circuit diagrams much easier to read. You will think so, too, once you get the idea.

87

## LATE AMERICAN SIGN LANGUAGE

The various symbols used in electronic circuit diagrams are a kind of shorthand. They are somewhat like stick drawings of men. Key elements are illustrated in the simplest possible way—just as a stick drawing of a man reduces him to a few lines and circles.

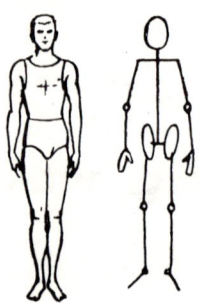

### ANTENNAS

Now, let's do the same with an electronic element, an antenna. To do it, we have to go back a few years, to the time when transmitting antennas looked like this—built with wooden spreaders holding a number of wires apart. All wires were tied together into one lead-in.

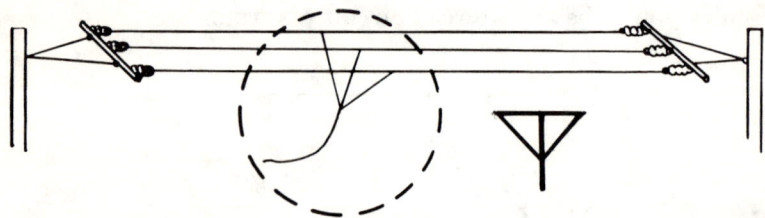

Nobody wanted to draw a whole antenna. So a key element, the triangular feed-line connection, was chosen to indicate an antenna.

All electronic circuit elements are made up in this fashion. A few lines represent, in a crude way, a certain electronic part. Let's get acquainted with some more of them.

## GROUNDS

Originally, just about every electronic device had to have an outside ground like the one on the Two Hour radio. In those days of insensitive equipment, one could neither receive nor transmit very far without a ground. (Ships at sea depended upon crystal receivers less sensitive than the Two Hour radio. Their antennas were whoppers—and the ocean was the ground!)

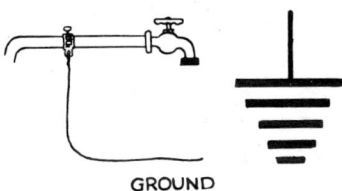

GROUND

As equipment improved, an outside ground was no longer required. However, many parts are connected to a common point, which serves as a kind of ground. Hence, you will find the ground symbol in most circuits. It may indicate connection to a metal chassis or common bus lead, or perhaps to several common points or lugs all wired together.

## WIRES

So much for antennas and grounds. Now let's look at the wires which tie the various parts together.

WIRE LEADS

Remembering the sign-language idea, you can see why a wire is simply a straight line. In a typical schematic diagram, most lines either run vertically or horizontally—for no good reason except that they look better and are easier to draw on a drawing board than lines which dive off at some angle.

When we want to hook two lines together, we need to indicate this. Again we do it in the simplest way possible—with a dot.

In any schematic, one line must almost always cross over another; so we must also indicate *no connection*. Unfortunately, diagram makers have never quite gotten together on this one. Hence, there are various systems. One is to simply

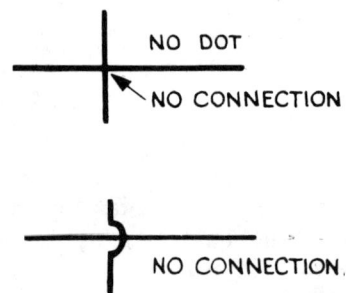

cross the lines, leaving off the dot. The other is to put a small hump in one of the lines. The latter, although a bit clearer, takes more time on the drafting board, which probably accounts for the fact that the first system seems to be used more. Sometimes a dot is not used at all. In this system, a connection is indicated by the lines crossing over, and a hooked line indicates no connection.

## CARBON RESISTORS

Now, let's look at some actual parts. One of the most common is the resistor. Again, a simple kind of sign language was developed. A zigzag path offers "resistance" to movement. Since a resistor offers resistance to electrons, we use a zigzag line to indicate a resistor.

Resistors come in two basic types—carbon and wirewound. Both serve much the same purpose, the difference being in the wattage they can handle without burning up. In transistor circuits, you will work mostly with carbon resistors because the current is low in most circuits. Such resistors are rated in wattage ($\frac{1}{4}$ watt, $\frac{1}{2}$ watt, etc.) as well as in resistance.

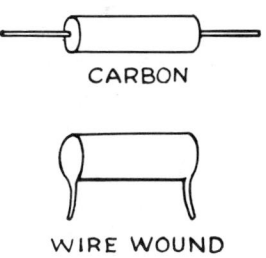

Resistors have always been tiny devices and—like everything else in electronics—are getting smaller all the time. So, the manufacturers have had to devise some way of indicating their resistance values, *without* depending on letters or numbers. Even if the printing were big enough to be readable, it usually will burn off the first time the resistor warms up a little. This makes it tough for the service technician to replace the resistor, because he can't tell what its resistance value should be merely by looking at it.

The answer is a system of color coding, with bands of color running around the resistor. (Some resistors have the value stamped on them, along with the color code.) The system has undergone some modification from time to time. Hence, in an old radio you may find a resistor which does not conform to the present method. But the parts you buy today, or find in any sets built within the past few years, will use the same system.

Actually, the system is simple. Each band indicates a number (or string of zeros), depending on the color. The first three

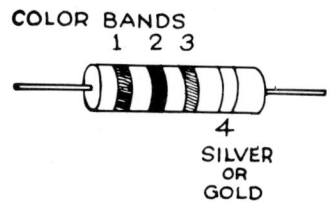

bands may be any of nine different colors. The fourth band (if used) ordinarily is silver or gold. In "decoding" a resistor, the figure is read *toward* the silver or gold band. If a fourth band is not used, the figure is read from the band closest to one end of the resistor.

Here is a table of the colors, plus the numbers they indicate:

| Color | First Band | Second Band | Third Band |
|---|---|---|---|
| Black | — | 0 | — |
| Brown | 1 | 1 | 0 |
| Red | 2 | 2 | 00 |
| Orange | 3 | 3 | 000 |
| Yellow | 4 | 4 | 0,000 |
| Green | 5 | 5 | 00,000 |
| Blue | 6 | 6 | 000,000 |
| Violet | 7 | 7 | 0,000,000 |
| Gray | 8 | 8 | 00,000,000 |
| White | 9 | 9 | 000,000,000 |

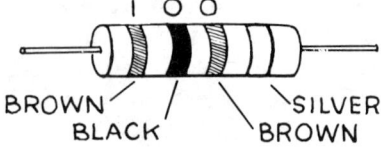

Let's take some examples. Suppose our resistor has a brown first band, a black second band, and a brown third band. Referring to our table, we find that the resistance is 100 ohms.

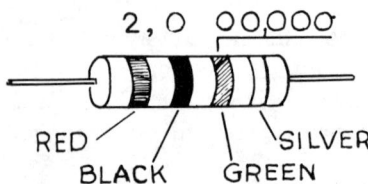

See how it works? Let's try another case. This time we have a red first band, black second band, and green third band. Add a comma and it will be a bit clearer—2,000,000 ohms. This is a common value. However, to save space, it is usually written *2 megohms*. (*M*egohm means *million* ohms.) It is often abbreviated still further—for example, 2 meg. Sometimes fractions are used: ½ meg means one-half million, or 500,000, ohms.

*Values* of 1000 ohms or more, but less than 1,000,000 ohms, are often written with the symbol $K$ ($K$ means 1000). Thus, a 1K resistor is a 1000-ohm resistor. A 500,000-ohm resistor may be written as either 500K or ½ meg.

What about the fourth band on a resistor? This silver or gold band (sometimes omitted) indicates the *tolerance* of the resistor. The resistors you will be working with usually will have a silver band, which means a tolerance of 10%. Thus, a 1000-ohm (1K) resistor with a silver end-band is within 10% *plus* (1000 plus 100, or 1100 ohms) or 10% *minus* (1000 minus 100, or 900 ohms). So, a 1000-ohm resistor with a silver end-band may actually measure anywhere from 900 to 1100 ohms. This is close enough in most circuits.

In key spots in television sets, gold-band resistors with a tolerance of 5% may have to be used. Resistors with no tolerance band need not be avoided. The absence of a fourth band merely means the tolerance is 20%.

There is a quick and easy way to decode resistor values. This is to utilize a low cost color code guide like those shown in the photo. These devices have small cardboard wheels which can be set up to match the colors of the bands on the resistors. When this is done, the resistor values appear automatically.

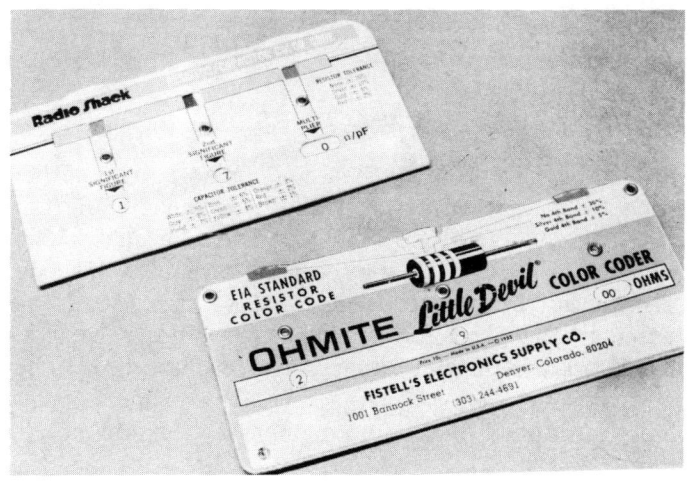

## WIREWOUND AND VARIABLE RESISTORS

Since wirewound resistors are usually much bigger than carbon resistors, their values are stamped right on the resistor, or on the carton. The printing on the resistor may burn off, of course, should the resistor become too hot. When this happens, the service technician hauls out a Howard W. Sams PHOTOFACT diagram and finds out from it what the part should be. You can see what a fix he would be in if he *couldn't* read a schematic!

So far, all we have talked about are nonadjustable items— once in the set, they retain (we hope!) the same value. Of course, there are also adjustable components, usually capacitors and resistors. The volume control on a radio is a good example of a variable resistor. Note that we indicate it with the regular resistor symbol, and then use an arrow to show that it is adjustable. The arrow technique is used on other parts, too, as you will see later.

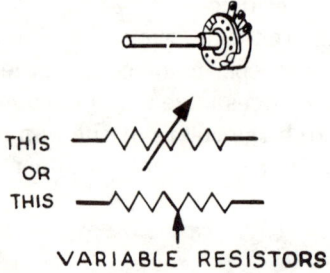

VARIABLE RESISTORS

## CAPACITORS

Capacitors (sometimes called *condensers*) are another key building block in electronic circuits. These components come in many shapes and sizes. The more common ones are shown in the drawings. Some, such as ceramic and mica, for example, sometimes can be used interchangeably; this fact is usually indicated in the parts list which accompanies a construction article. Sometimes Mylar capacitors are preferred. Paper capacitors are not as versatile as ceramic and mica types, and electrolytic capacitors are used only for certain specialized jobs.

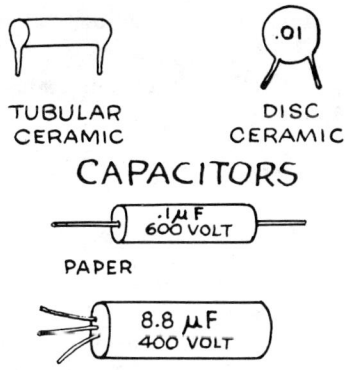

## CAPACITORS

All capacitors are made up of "plates" separated by a layer of insulating material (which may even be air). Hence, the symbol for a capacitor is two lines separated by space. Should the capacitor be an electrolytic, it will have polarity, like a battery. The polarity is usually shown on the schematic symbol.

As you will discover in looking at circuit diagrams, capacitors may also be shown with one line slightly curved. The curved line indicates the side of the capacitor going to the grounded side of the circuit.

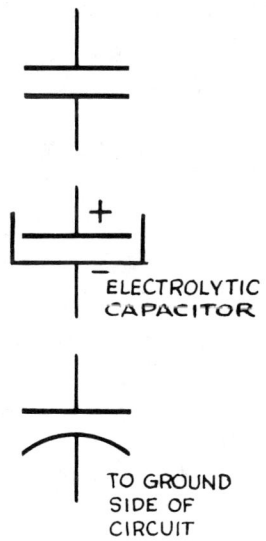

At one time, capacitors were color coded. However, the system was so confusing that, much to everybody's relief, *most* (but unfortunately not *all*) capacitors today have the value stamped right on them. Since capacitors seldom heat up much (except electrolytics, on occasion), the markings stay on capacitors fairly well.

## ONE MILLIONTH OF A MILLIONTH

The value (electrical size) of a capacitor is not simple to show because, as with resistors, there is more than one way to indicate the same size. The first consideration is voltage rating, which is easy—for example, a capacitor may be rated at 15 volts for use in transistor circuits. This rating is the maximum dc voltage the capacitor can withstand without breaking down. For this reason, it is essential to use a capacitor with a voltage rating *higher* than is likely to be encountered in the set. In a transistor set with 9 volts, the capacitors will often be rated at 12 or 15 volts—just to allow a safety margin.

Now for the capacitance value, the microfarad. Actually, the basic unit is the farad. However, a capacitor with a capacitance of one farad might be as big as a bathtub! So, the largest capacitor you are likely to encounter will be rated in *micro*farads. A microfarad (abbreviated $\mu$F) is one millionth of a farad.

Electrolytic capacitors are usually one or more microfarads. With paper or ceramic capacitors, the usual rating is a decimal fraction of a microfarad. A common size is 0.01 $\mu$F. Everything would be just fine if all capacitors were rated in this way. As they get smaller, however, a new system takes over.

For example, a common size of ceramic or mica capacitor is 0.00025 $\mu$F. This value may also be expressed in another way—in terms of units of one millionth of one millionth of a farad—pretty small indeed! These units are called *picofarads* (abbreviated pF).

To reduce a parts value like 0.00025 $\mu$F to picofarads, we must multiply it by one million. If you remember how to work decimals, the job is easy. For example:

$$0.00025 \times 1{,}000{,}000 = 250 \text{ pF}$$

You can also go the other way. To translate 500 pF into microfarads, we simply divide by 1,000,000. For example:

$$500 \div 1,000,000 = 0.0005 \ \mu F$$

If that seemed complicated, don't despair. The reason is that you probably don't use decimals very often. Fortunately, you can quickly learn to do the job mentally by adding zeros. Hence, to go from 0.00025 $\mu$F to 250 pF, simply move the decimal point six places to the *right*; to go from 250 pF to 0.00025 $\mu$F, move the decimal point six places to the *left*. Or—and easier—do the job on a calculator!

As a last resort, you can always show your shopping list to the counterman at the radio parts distributor. He has to understand the markings in order to sell capacitors!

### VARIABLE CAPACITORS

Just as with resistors, an arrow indicates a variable capacitor. It may be drawn through the capacitor symbol, or the curved plate may be terminated with an arrow. Tuning capacitors are usually packed in a box, with the size stamped on the side of the box. Since they are almost always labeled in picofarads, no translating job is needed. Note that the rotor is the set of *moving* plates and the stator is the set of *stationary* plates.

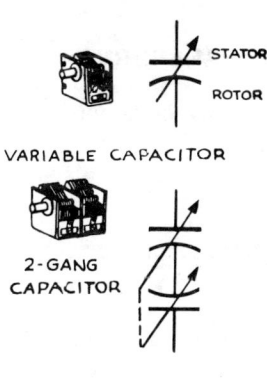

VARIABLE CAPACITOR

2-GANG CAPACITOR

### COILS

Coils are easy to indicate in a schematic—you simply draw what looks like a pig's tail. A variable coil is indicated by an arrow. If the coil is tapped, the tapped connection is also shown. Three parallel lines indicate the core.

The unit for rating a coil is the *henry*. Except for audio and filter chokes, however, the coils and rf chokes (a specialized type of coil) you encounter will be rated mostly in millihenries or microhenries. The box is usually stamped with the value, so

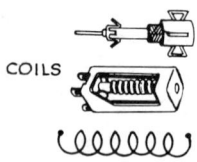

COILS

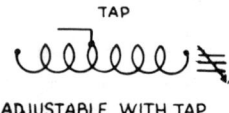

TAP

ADJUSTABLE WITH TAP

ordering parts is easy. Be certain never to mix up the *milli* and the *micro*—the difference is very large indeed. A millihenry is equal to one-thousandth, and a microhenry to one-millionth, of a henry.

## TRANSFORMERS

Transformers are simply coils with cores, usually of iron. The core is indicated by lines, as shown in the drawing.

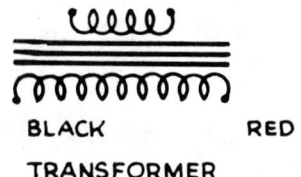

BLACK        RED

TRANSFORMER

Transformers often have a number of windings, identified by color coding the various leads. This color code is very important—follow it carefully if the text so advises.

## DIODES AND TRANSISTORS

Transistors are so numerous and complex that to even try to indicate the possible types which are (or will be) available is impossible. The diagrams illustrate the two types found in this book. Notice that the symbols for the npn and pnp types differ only in the direction the arrows point.

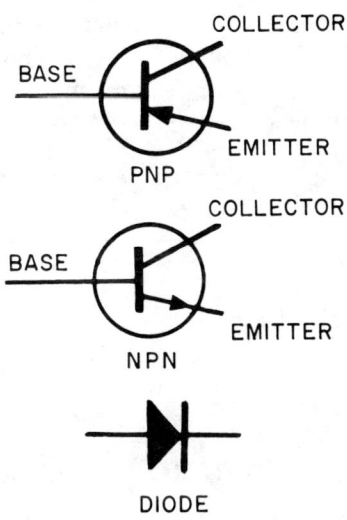

Sometimes the manufacturer changes the arrangement of the leads after a transistor has been in production for awhile. If the transistor you buy doesn't have the same terminal arrangement as the one in this book, check it carefully against the base diagram (usually packed with the transistor). If the matter *still* isn't entirely clear, see a transistor manual. (Such manuals are issued from time to time by transistor manufacturers.) The idea, of course, is to be able to identify the *collector, base,* and *emitter* leads.

## FET UNITS

The next generation of transistors after the npn and pnp designs are called FETs—Field Effect Transistors. The elements are different from those of other transistors; the leads are identified as "S," "D," and "G" for *source, drain,* and *gate.*

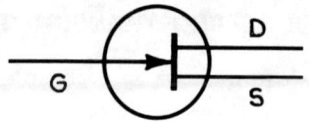

## INTEGRATED CIRCUITS

We have already met ICs in previous chapters. As mentioned, an IC is a wonderfully compact combination of transistors, diodes, and other "parts." The symbols vary, usually being a simple square, rectangle, or triangle with the leads numbered. Two examples of IC symbols are shown.

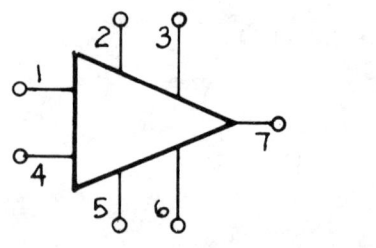
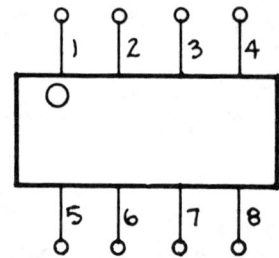

## SWITCHES

Switches are easily indicated, too. A switch often has multiple sections, all acting at the same time when the switch is turned. A multiple-section switch is shown. Notice how the units are tied together with a dash line, indicating that they are all operated by the same button or from the same shaft. The same technique is used with variable capacitors and resistors, for example, in which several components are controlled by the same shaft.

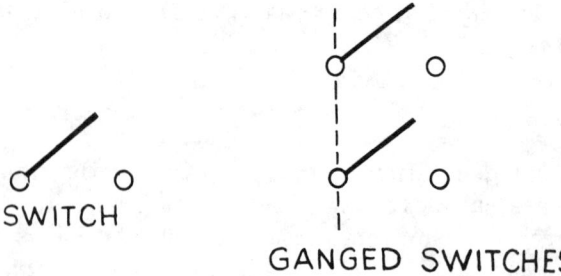

## PHONES

The symbol for a headphone is one of the simplest of all, particularly for only a single phone. The symbol for a pair of phones is almost as simple.

SINGLE PHONE        DOUBLE PHONE

## TUBES

Although the sets in this book do not use tubes, you will want to be able to recognize a tube symbol when one comes your way. Tubes are of many different designs. Tube manuals give the type numbers, characteristics, schematic symbols, and base connections of the tubes.

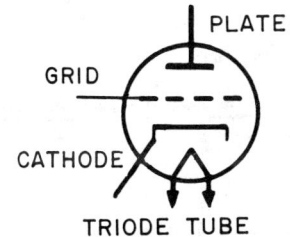

TRIODE TUBE

## OTHER COMMON PARTS

Once you have become acquainted with the appearance of the various parts and their symbols, your next step is to fit the

SPEAKER      BATTERY      PHONE JACK

pieces together and apply this sign language in drawing and reading actual circuit diagrams. This we will do in the following chapter.

**CHAPTER 9**

# Reading Circuits

Once you understand the various symbols for electronic parts, you will have gone far toward being able to read circuits. In addition, there are certain tricks of the trade you will need to know.

As a starting point, let's consider the pictorial diagram for the Two Hour radio—about as simple as any electronic device

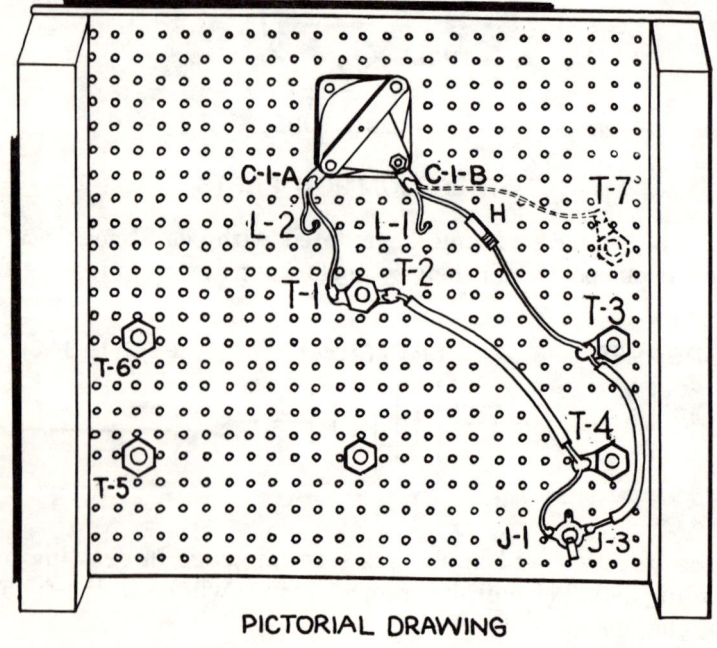

PICTORIAL DRAWING

you will ever encounter. First, take a look at the pictorial diagram of the underside of the chassis, just to become oriented.

Now we will overlay with the circuit symbols. An experienced builder *could* use the resulting circuit as shown for wiring. But he would be a lot more comfortable using the *same* circuit redrawn in the form of a *schematic diagram*.

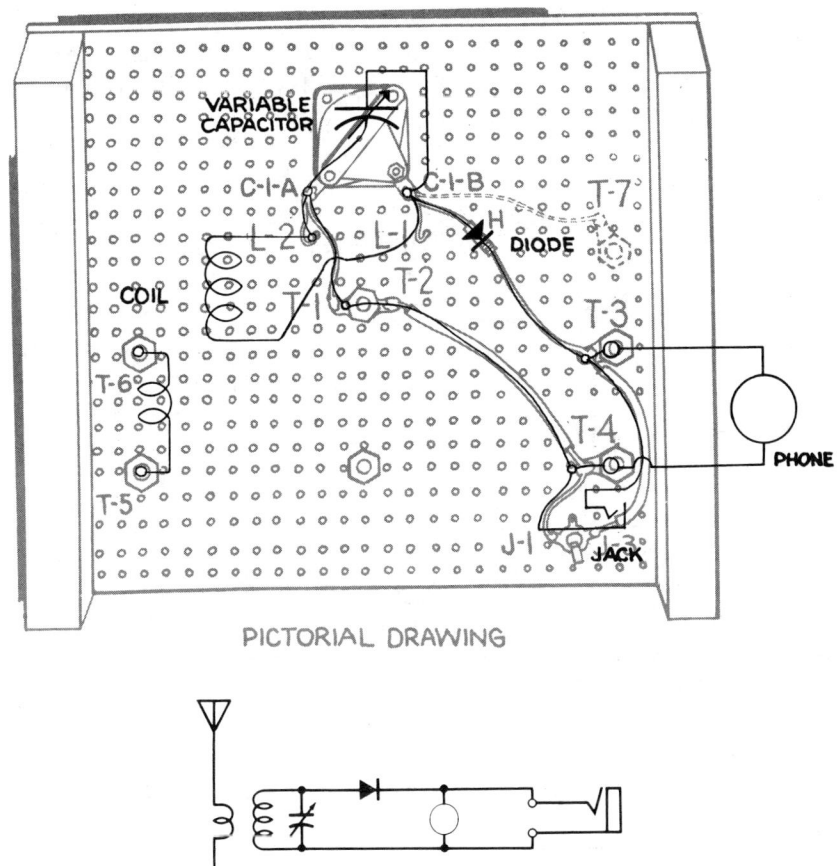

PICTORIAL DRAWING

SCHEMATIC DIAGRAM

At first glance, this new diagram may look quite different from the first one. However, if you will trace out the wiring, lead for lead, you will discover that it is exactly the same electrically, the difference being simply in the arrangement of parts and wire leads.

## READ FROM LEFT TO RIGHT

Reading circuits on complicated equipment (for example, television sets) could be a pretty formidable job if there weren't some sort of pattern to follow. Such a pattern was arrived at many years ago; essentially, it is just like reading. You mentally follow the radio signal through the diagram, your eyes moving from left to right just as they do in reading this page.

In our Two Hour radio, the signal enters the antenna, goes to the coil, then to the diode where it is detected, and finally to the earphone. In the diagram, therefore, we lay out the parts so that we can follow the signal path from left to right.

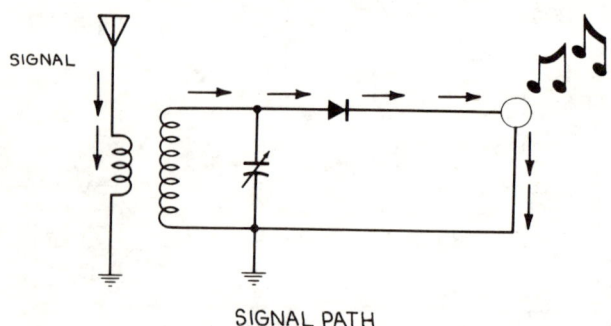

SIGNAL PATH

See the diagram that shows the signal path. The signal picked up by the antenna goes to the smaller of the coils, where, as it passes through, it "induces" a current in the larger coil. From the coil, the signal current passes to the diode detector, which transforms it into an audio signal we can hear.

In such a simple circuit, it is easy to visualize what is happening. With complex circuits, the task is more difficult, of course, but as you work with equipment you will become more and more skilled at mentally tracing a signal through a circuit diagram. However, we have made progress if you will remember that the left side of a circuit is the *input* and the right side is the *output*.

Some other things are often puzzling to the beginner. For example, *within* the wiring of a set there often are several points where wires can be joined without upsetting the circuit in any way. Usually, the reason for making connections in a given fashion is purely mechanical convenience. In some radio-frequency circuits—or even in high-gain audio hookups—the length of wire leads is important. But generally a lot of flexibility is possible.

Let's take a look at the Selective AM Tuner as an example of how this works out. Both of the variable capacitors have the rotors (moveable plates) connected to the "ground" side of the circuit. Terminal C-1-B goes first to clip O-4 (no circuit reason—simply to avoid overheating C-1-B), and from there a lead goes to V-4, where it ties into a string of clips, the wiring finally terminating in T-1 after passing through clip F-22.

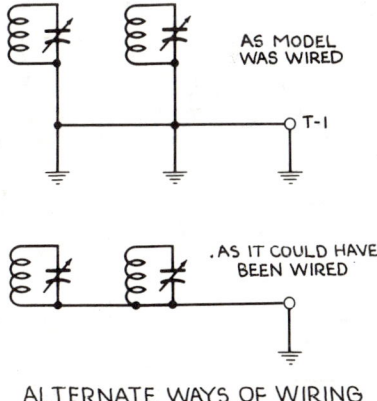

ALTERNATE WAYS OF WIRING
SELECTIVE AM TUNER

However, the lead from C-1-B *could* have gone directly to C-2-B. Electrically, the result is exactly the same. But there were two other considerations. Most important, making one connection to C-2-B without endangering the plastic leaves between plates requires care—two connections multiplies the problem. Second, it is always good practice to have a grounding surface surrounding the balance of the wiring if possible. Printed circuit boards are invariably set up this way. Admittedly this grounding technique is of small consequence with a low-gain circuit like the Selective AM Tuner, but we might as well start out with good practice in mind. The circuit diagrams shown on page 105 illustrate both approaches.

The connections to L-1-B are another example. Both the

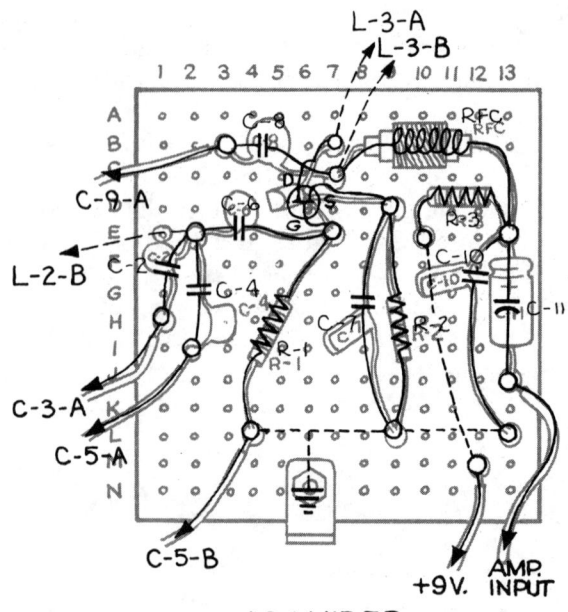

AS WIRED

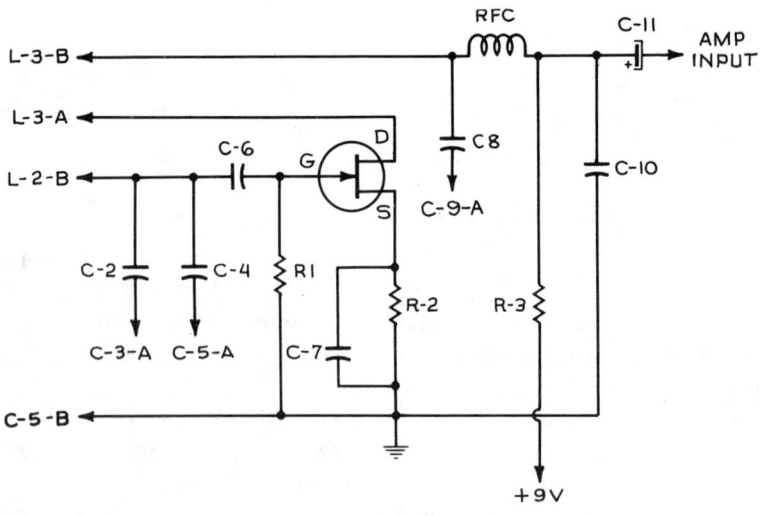

CIRCUIT DIAGRAM

diode and the lead from C-2-A *could* have gone directly to L-1-B, omitting clip F-16. But this would have meant having two soldered connections to L-1-B, and coils, like the plastic-leaf capacitors, utilize such tiny wire that they are delicate beasts at best, and overheating is to be avoided.

As a next step, let's look at circuit diagrams for the detector board of the shortwave set in Chapter 12, as it *is* wired and as the circuit is drawn following standard practice. (See page 106.) Mechanically, they are similar but *not* the same. *Electrically* they are *identical*, but the circuit diagram is much easier to follow.

Circuits containing ICs, because so many parts are included in such a compact space, tend to allow the layout and the circuit diagram to look more alike.

## APPLYING WHAT YOU HAVE LEARNED

From the preceding examples, you should have a fairly good idea of what an electronic circuit diagram is all about. Now go back and look at the circuit diagrams which have been included in each chapter, and locate the various parts on both the pictorial illustration and the circuit diagram

Each diagram you check out will seem easier—suddenly you will realize that you are reading circuits, and not depending upon pictures. And you will have acquired a skill that can be useful in many fields, from ham radio for fun to computers as a livelihood.

**CHAPTER 10**

# Utility Amplifier — First Step to Stereo

A highly useful device for any electronics experimenter is an audio amplifier. For example, it can be used with a variety of tuners, like those described in previous chapters, to raise the output to loudspeaker level. Likewise, it can serve as a phono amplifier for a record changer. And, as will be covered later, two such units provide a way to create a simple stereo system, a system a bit unique because it will operate from two small batteries. In addition, as you learn more about electronics, you may want to fit the amplifier with a couple of probes and create a "signal tracer" for troubleshooting other electronic gear.

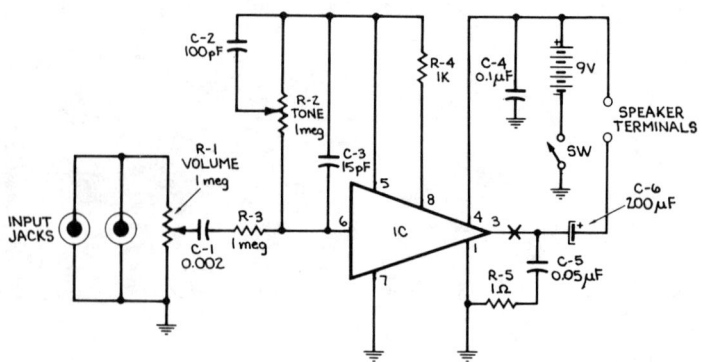

## BUILT AROUND ONE IC

The unit to be described utilizes a ½-watt integrated circuit which contains 13 transistors, 5 diodes, and 9 resistors—all in

### Shopping List

| Quantity | Description | Labeled in Diagrams |
|---|---|---|
| 1 | 1 megohm variable resistor with switch | R-1 |
| 1 | 1 megohm variable resistor | R-2 |
| 1 | 1 megohm, ¼ watt resistor | R-3 |
| 1 | 1000 ohm (1K), ¼ watt resistor | R-4 |
| 1 | 1 ohm, ¼ watt resistor | R-5 |
| 1 | 0.002 µF, 100 volt (or higher) Mylar or ceramic capacitor | C-1 |
| 1 | 100 pF mica capacitor | C-2 |
| 1 | 15 pF mica capacitor | C-3 |
| 1 | 0.1 µF, 100 volt (or higher) Mylar or ceramic capacitor | C-4 |
| 1 | 0.05 µF, 100 volt (or higher) Mylar or ceramic capacitor | C-5 |
| 1 | 200 µF, 15 volt (or higher) electrolytic capacitor | C-6 |
| 1 | Audio amplifier IC (Motorola 1306P or Sylvania ECG745 or equivalent) | IC |
| 1 | Socket for IC | |
| 1 | Soldering-lug terminal strip with one non-grounded lug | TS |
| 1 | Screw-type terminal strip with two terminals | TL |
| 2 | Solder lugs | |
| 2 | Phono jacks | PH |
| 1 | 9 volt transistor battery | |
| 1 | Battery connector | |
| 1 | Battery holder | BH |
| 1 | Perforated board, 2¾ × 2¾ inches (7 × 7 cm) | |
| 37 | Push-in terminals for board | |
| 2 | Metal spacers (see text) | |
| 2 | Knobs | |
| 1 | Chassis box, 5 × 7 × 2 inches (12.7 × 17.8 × 5.1 cm) | |
| As needed | Shielded wire, bare hookup wire, insulated hookup wire, two-conductor speaker wire, bolts and nuts. | |

a tiny 8-pin "package." With only a few additional parts, this IC can be transformed into an amplifier and preamplifier equal in performance to a unit which a few years ago would have required many, many components and a major wiring job. The amplifier is built utilizing the perforated-board construction method with which you are already familiar.

## WIRING THE BOARD

Insert push-in terminals on the top of the board as follows:

| Row | Holes |
|---|---|
| A | 6, 7, 13 |
| C | 7, 10 |
| D | 1, 2 |
| E | 10, 12 |
| F | 5, 9 |
| G | 5, 9, 13 |
| H | 1, 4, 5, 9, 10, 12 |
| I | 5, 9, 11, 12 |
| J | 2, 5, 10, 12 |
| K | 1 |
| L | 6, 12 |
| M | 5, 6, 12 |
| N | 1, 5, 13 |

On the bottom of the board, mount soldering lug terminals in enlarged holes B-2 and M-10, using 6-32 machine screws and nuts. These bolts will also be used to hold the board to the chassis, so they will need to be long enough to accommodate the metal spacers mentioned later.

Follow the drawing, and, with bare wire, connect the various push-in terminals and soldering lugs together on the bottom of the board.

Soldering the parts in place on top of the board is easy—with one exception, wiring in the 8-pin socket. This will be a test of your soldering skill. For details, see Chapter 5 and follow the instructions given there. Solder the socket to clips F-5, F-9, G-5, G-9, H-5, H-9, I-5, and I-9. Be very careful in soldering so that you don't *desolder* the connections to the socket. This *can* happen; in one model built by the author, he failed

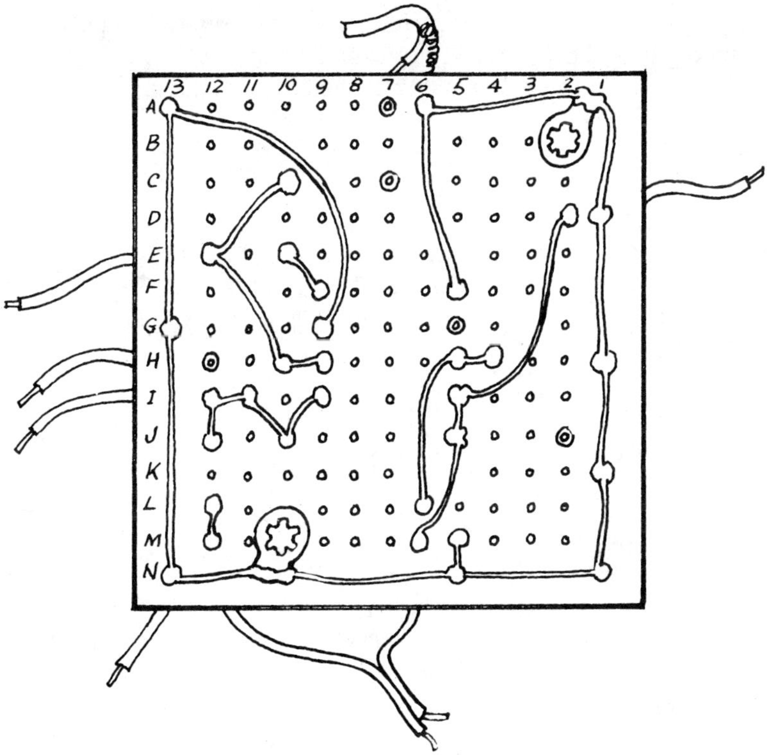

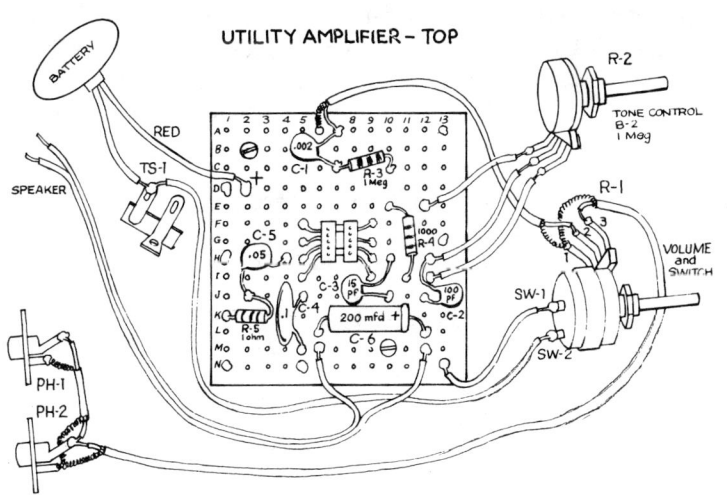

UTILITY AMPLIFIER – TOP

to notice that pin 1 had come loose. As a result, when power was applied the IC quickly overheated and was destroyed.

Now solder the other parts in place, as follows:

| Part | Between |
|------|---------|
| R-3  | C-7 and C-10 |
| R-4  | E-10 and I-11 |
| R-5  | K-1 and J-2 |
| C-5  | J-2 and H-4 |
| C-1  | A-7 and C-7 |
| C-4  | J-5 and M-5 |
| C-6  | L-6 and L-12 (+ end to L-12) |
| C-3  | H-10 and J-10 |
| C-2  | J-12 and H-12 |

## AUXILIARY PARTS

The foregoing completes the installation of the parts on the board. Now, we will add auxiliary parts, and, as was done with a previous unit, make certain our unit is operating *before* we install it in the metal cabinet.

By following the drawing, add the battery connector with the red lead going to D-2 on the board and the black lead to terminal strip point TS-1. From TS-1, run a lead to SW-2 on the volume control (potentiometer). Connect SW-1 to N-13.

The leads that connect to the volume control and the phono jacks all utilize small shielded wire. The photo shows how to pull the center lead through the shield so that the shield becomes one lead and the center wire becomes the other lead.

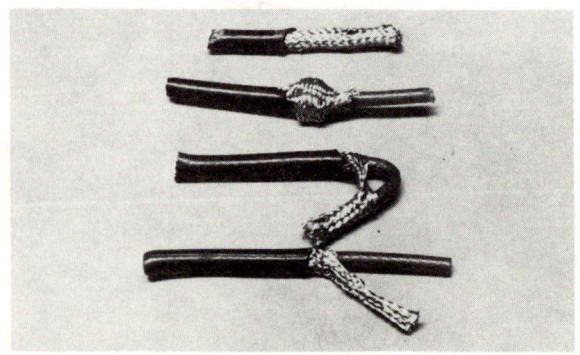

The two phono jacks, PH-1 and PH-2, are connected together with a 4-inch length of shielded wire as shown in the drawings. From PH-2, run a shielded wire to R-1 with the shield connected to R-1-1 and the center lead connected to R-1-3. From R-1-2, the center lead of a shielded wire goes to A-7, and the shield goes from R-1-1 to A-6. Connect the tone control, potentiometer R-2, to E-12, H-12, and I-12 as shown in the drawing.

The only remaining wires are the twin-conductor speaker leads. These connect to M-6 and M-12.

Now, check and recheck the wiring.

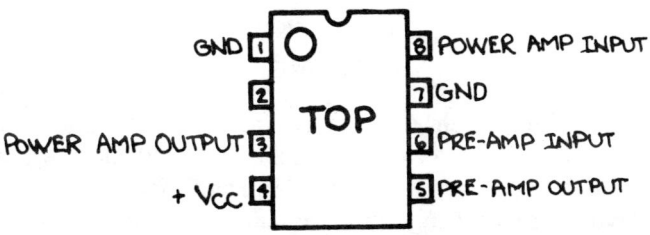

Next, with the switch off, plug in the IC. Note that the device has a small circle on the top of the case. Be certain that you plug the IC into the socket so that this circle is closest to the socket lead which connects to F-5. If you install the IC backward, it will probably be ruined quickly when power is applied. Also, be careful in inserting the IC into the socket: make certain that *all* of the pins go in and that none of them "misses" the hole.

## ADD A SIGNAL

The next step is to provide some kind of signal input. This can come from a record changer (ceramic cartridge) like that shown in the photo, or from any tuner with enough output to operate an earphone. The Two Hour radio, Selective AM tuner, Transistor AM receiver, or Hi-Fi AM tuner are all possibilities. Make up a shielded cable and plug as shown in Chapter 7 if you are connecting to one of the tuners.

Connect an 8-ohm speaker to the speaker leads.

As the final step before testing, snap the 9-volt battery in place.

## TESTING

Turn on the switch (on the volume control), and advance the control. If all is working properly, you will hear whatever signal the tuner or phono player is providing—greatly amplified. If you hear nothing, shut the amplifier off quickly. Touch the leads from a crystal earphone to A-6 and C-7 to make certain a signal is going *into* the unit. If there *is* signal input, you know that the amplifier is not working. The next step, of course, is to recheck all wiring and *all* soldered connections, with special attention to the tricky soldered connections to the IC.

## FINISHING UP

Once you have the unit operating properly on the bench, the final step is to mount it in the chassis box. Drill holes so that terminal-strip lugs TL-1 and TL-2 do not touch the chassis. Then bolt the strip to the chassis, and solder the speaker leads to the lugs. The phono jacks are mounted in suitable holes, again with bolts and nuts.

The battery holder is bolted to the chassis. The board is mounted on two small metal spacers long enough (and with bolts long enough) so that the board and all the push-in terminals clear the chassis. These bolts go through the solder lugs and holes at B-2 and M-10. Two holes drilled in the front of the chassis allow the tone control and volume control to be secured with the nuts that are on the shafts. The small

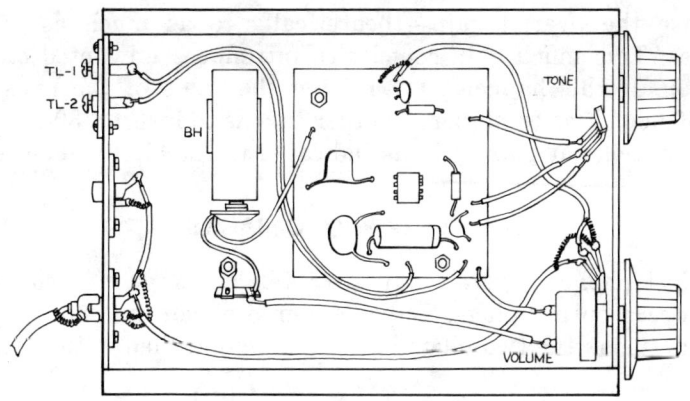

spacers can be obtained from a radio parts supplier or you can use ½-inch lengths of small copper or aluminum tubing; the inside diameter of the tubing must be large enough for the bolts to pass through.

Slide on the knobs; tighten the set screws. We're done!

The amplifier as designed is best suited for amplifying a ceramic-pickup phono player or other source with fairly high output. It will work, of course, with the lower output of a tuner. But if you want to use it primarily with a tuner, particularly one of limited output, replace 1-megohm resistor R-3 with a resistor of 10,000 ohms (10K). This will greatly in-

crease the overall gain—theoretically to as much as 1000 times. This much gain sometimes introduces unwanted oscillation. If this happens, try raising the value of the resistor to 50,000 ohms or higher. Another cure is to insert a 30-microhenry choke at point "X" as indicated in the circuit diagram.

## WHY TWO PHONO JACKS?

Most phono players today are intended for stereo, so they have two output plugs. To use a stereo player on "mono," we simply hook the two plugs in parallel by paralleling the jacks.

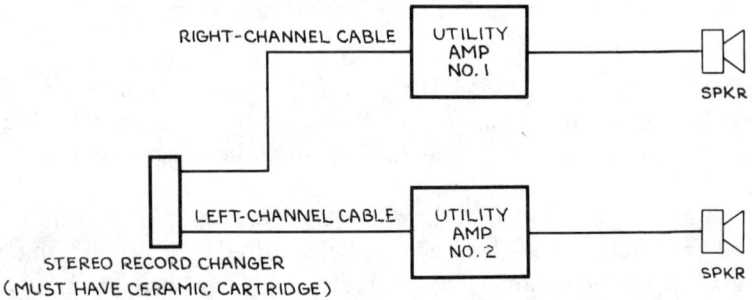

But yes, we *can* use the utility amplifier for stereo. We simply use *two* amplifiers, as shown in the block diagram. Such a hookup won't rattle the windows, but it will give you honest-to-goodness stereo. And the auto speaker and "hang up" enclosure will yield quality which will probably surprise you for such a low power and inexpensive setup.

CHAPTER 11

# Two Tuners: CB and Aircraft

All of the tuners described up to this point have been for the standard am band. Of course, this is only one band in the radio spectrum; there are many others filled with interesting signals. By teaming up a simple tuner with either the Loudspeaker Amplifier or the Utility Amplifier described previously, we can create a receiver capable of tuning other bands.

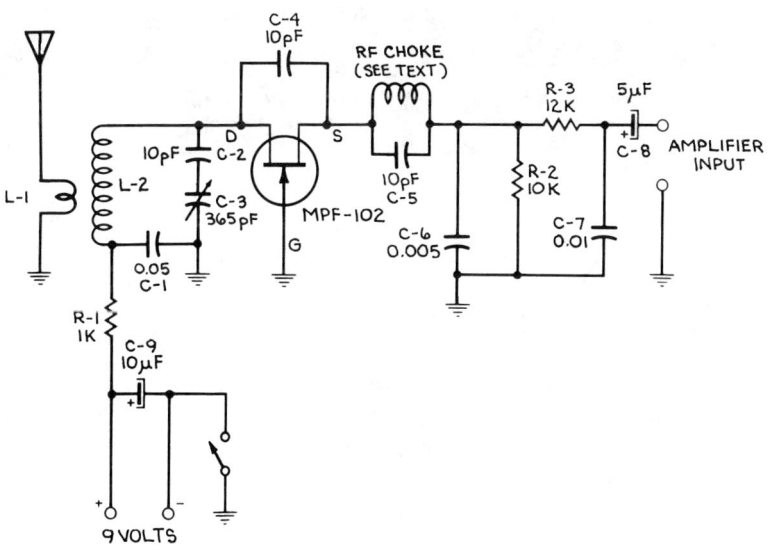

117

## Shopping List

| Quantity | Description | Labeled in Diagrams |
|---|---|---|
| 1 | 0.05 µF, 100 volt Mylar capacitor | C-1 |
| 3 | 10 pF mica capacitor | C-2, C-4, C-5 |
| 1 | 365 pF variable capacitor (Calectro A-1-232 or equivalent) | C-3 |
| 1 | 0.005 µF, 100 volt Mylar capacitor | C-6 |
| 1 | 0.01 µF, 100 volt Mylar capacitor | C-7 |
| 1 | 5 µF, 16 volt electrolytic capacitor | C-8 |
| 1 | 10 µF, 16 volt electrolytic capacitor | C-9 |
| 1 | 1000 ohm (1K), ¼ watt resistor | R-1 |
| 1 | 10,000 ohm (10K), ¼ watt resistor | R-2 |
| 1 | 12,000 ohm (12K), ¼ watt resistor | R-3 |
| 1 | Coil assembly (see text) | L-1, L-2 |
| 1 | Transistor (FET), Type MPF-102 or equivalent | |
| 1 | Rf choke (see text) | RFC |
| 1 | Toggle switch, spst | SW |
| 1 | 9 volt battery | |
| 1 | Battery connector | |
| 4 | Soldering lugs | |
| 2 | 6-32 bolts with nuts (length to suit spacers) | |
| 2 | Spacers (see text) | |
| 1 | Perforated board, 1¾ × 3 inches (4.4 × 7.6 cm) | |
| 18 | Push-in terminals for board | |
| 1 | Box (plastic with aluminum panel), approximately 5¹/₁₆ × 2⅝ × 1⅝ inches (12.9 × 6.7 × 4.1 cm) | |
| | Insulated hookup wire | |
| | Bare hookup wire | |
| | Shielded wire (Belden R-216 or equivalent) | |
| 1 | Knob for C-3 | |
| 1 | Phono plug | |

The tuner described in this chapter utilizes a superregenerative circuit, which is the simplest approach to tuning signals in the range 27 MHz to 150 MHz. Not only is the circuit simple, but the unit can be utilized on a number of different bands, simply by changing the tuning coil and the homemade rf choke.

For this chapter we have chosen two different bands: CB and aircraft. There are CB signals on the air virtually anywhere in the United States, and there is a certain fascination to listening to the CB chatter. Further, the broad-tuning superregenerator tends to pick up the strongest signal, which means that you can often listen for quite a while without doing much tuning.

If you live in a small town or a rural area, far from a major airport, the aircraft tuner is *not* an ideal project. The tuner will pick up aircraft for 50 miles or so if the planes are high in the air. However, the conversations on these frequencies are extremely brief, and unless there are a lot of them, you may have trouble finding one and tuning it in. If you *are* within range of a busy airport, you can pick up both the control tower and incoming planes, and all of this can make for interesting listening.

At any rate, build the CB tuner first and get it working. Later, you can convert it to an aircraft tuner by changing the coil and rf choke.

### INSERTING THE TERMINALS

Following the chart below, insert the push-in terminals into the top of the board. Then, temporarily bolt soldering-lug terminals at enlarged holes A-1 (bottom of board) and B-15 (top of board). The lug at I-14 will be added later.

| *Row* | *Hole Number* |
|---|---|
| A | 3, 7, 10, 14 |
| B | 8, 13 |
| C | 10, 13 |
| D | 10 |
| E | 8, 11 |
| H | 8, 11, 15 |
| I | 1, 7, 12, 14 |

### COILS AND CHOKES

As was mentioned earlier, the only difference between the CB tuner and the aircraft tuner is that the tuning coils and the rf chokes are of different values. The CB tuner, operating in the 27-MHz range, is less critical in construction. Because of this—plus the fact that there are more likely to be nearby CB stations than a busy airport within range—the CB tuner is the one to build first.

The tuning coil for the CB tuner can be wound on a small piece of plastic tubing that is just a whisker over 3/8 inch (9.5 mm) in outside diameter. The tubing used by the author came from the plastic case which is used to package an X-acto knife. Another approach is to wind the CB coil over a 3/8-inch form of some type (for example, the shank of a 3/8-inch drill bit covered with a layer of waxed paper) and apply several strips of household cement to hold the turns in place. After the cement dries, the coil will be self-supporting.

The coil in the original tuner used No. 24 wire. Somewhat larger or smaller wire could have been used. L-1 consists of 6 turns of wire; L-2 has 14 turns. The method of winding the coil is exactly the same as for the homemade coil for the Two Hour radio. Be certain to leave the lead from L-1 which goes to I-1 about 2 inches long so that it will reach I-1.

Like the coils, the rf chokes are homemade. The choke for the CB tuner consists of a 5-foot–3-inch length of No. 26 wire "scramble wound" on a 20K, 1-watt resistor. Solder the two ends of the wire to the resistor leads close to the resistor body. The resistor serves as a convenient coil form, and it provides leads for connecting the rfc into the circuit.

## ADDING PARTS AND WIRING

First mount the variable capacitor on top of the board in an enlarged hole at F-3. (There is a soldering lug under the nut that secures the tuning capacitor in hole F-3.) Then connect wires on the bottom of the board as follows:

- ☐ Solder lug at A-1 (S) to capacitor lug at F-3 (DS)
- ☐ Lug at F-3 (S) to I-7 (DS)
- ☐ I-7 (DS) to I-12 (DS)
- ☐ I-12 (S) to I-14 (DS)
- ☐ I-14 (S) to A-14 (DS)
- ☐ A-14 (S) to A-10 (DS)
- ☐ A-10 (DS) to C-10 (S)
- ☐ A-10 (S) to A-7 (S)

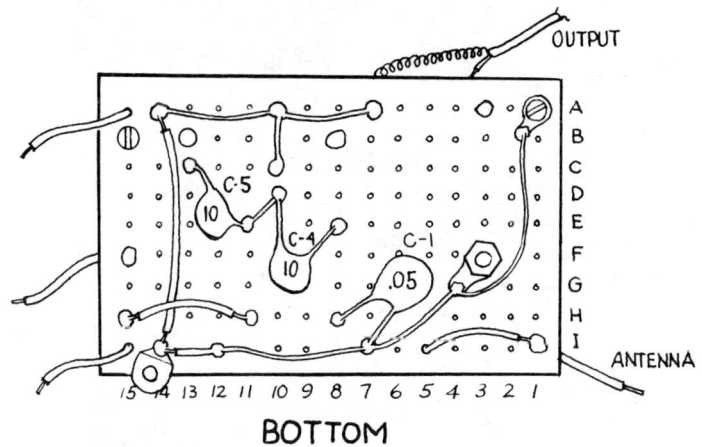

**BOTTOM**

- [ ] H-15 (S) to H-11 (S)
- [ ] E-11 (DS) to D-10 (DS)

Also on the bottom of the board, connect

- [ ] Capacitor C-5 from C-13 (S) to E-11 (S)
- [ ] Capacitor C-4 from D-10 (S) to E-8 (S)
- [ ] Capacitor C-1 from H-8 (S) to I-7 (S)

Now, solder the various parts in place on top of the board. Keep the leads as short as possible. (See page 122.)

- [ ] R-1 between H-8 (DS) and H-11
- [ ] C-9 between I-12 and H-15 ("plus" end to H-15)
- [ ] C-3-A (S) to I-7 (DS). Protect C-3-A from heat.
- [ ] L-2-B to E-8 (DS)
- [ ] L-2-A to H-8 (S)
- [ ] L-1-A to I-1 (DS)
- [ ] L-1-B to I-7 (S)
- [ ] C-2 between C-1-B (S) and E-8 (DS). Protect C-1-B from heat.
- [ ] C-8 between A-3 (DS) and B-8 (DS) ("plus" end to B-8)
- [ ] RFC between E-11 (S) and C-13 (DS)
- [ ] Connect C-13 (S) to B-13 (DS)
- [ ] R-3 between B-13 (DS) and B-8 (DS)
- [ ] R-2 between A-10 (S) and B-13 (DS)
- [ ] C-6 between B-13 (S) and A-14 (DS)
- [ ] Connect A-14 (S) to lug T-2

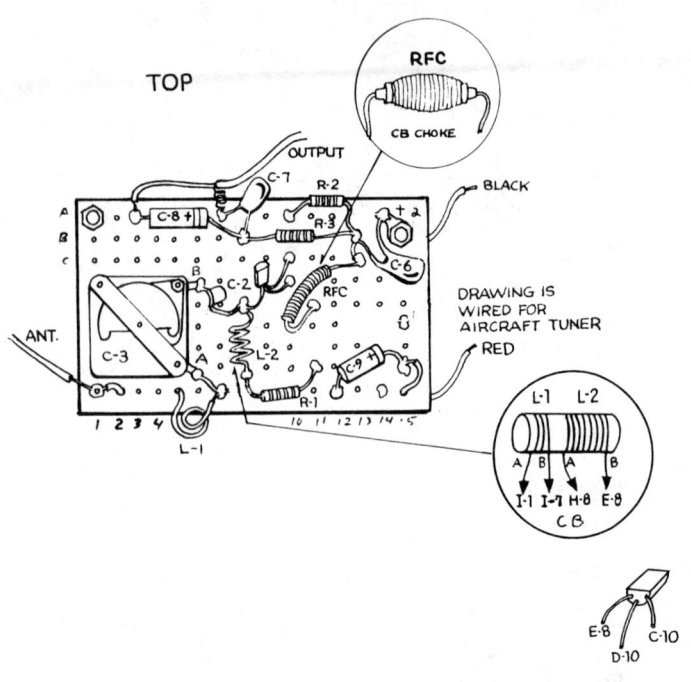

- [ ] Connect C-7 between A-7 (DS) and B-8 (S)
- [ ] Using the technique described earlier in the book to protect the transistor from damage, solder the transistor leads to D-10 and C-10 as shown in the drawing.
- [ ] Protecting the remaining transistor lead from heat as before, connect it to E-8 and solder in place. This connection joins C-2, L-2-B, and the transistor.
- [ ] Following the drawing shown in a later section ("Into the Cabinet"), add the battery connector and switch.
- [ ] Solder the antenna lead to I-1.
- [ ] Solder the inner wire of a small shielded cable to A-3, and solder the shield to A-7. The other end terminates in a phono plug. Secure the shielded cable to the board with a small twist of wire.

The board is mounted directly on the front panel by means of small metal spacers about ½ inch (13 mm) long. Short pieces of copper or aluminum tubing will do the job fine. A soldering lug is also bolted to the panel and soldered to I-14.

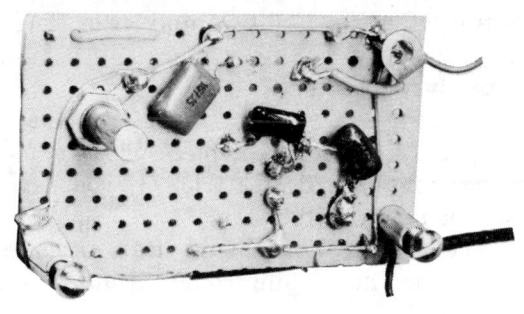

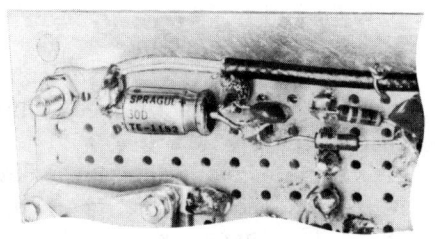

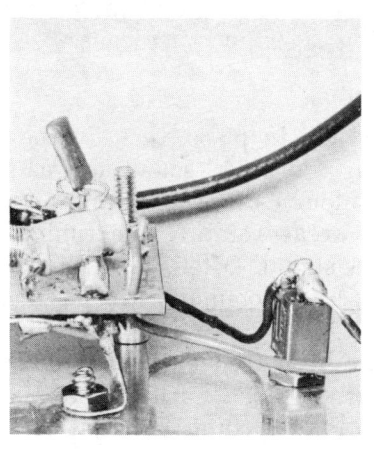

Plug the tuner into the Utility Amplifier or the Loudspeaker Amplifier.

Connect the 9-volt battery.

Connect some sort of antenna to the antenna lead. If you are in a busy CB area, a 9-foot piece of wire, as much of it vertical as possible, may do the trick. If you have an outside antenna, so much the better.

Turn on the switch. You should hear either a "rushing" sound or a signal. The rushing sound disappears as you tune in a CB station.

## CONVERTING TO AIRCRAFT

To convert the tuner to the aircraft band, we need to make a new coil and rf choke.

The choke is wound with 32 turns of No. 26 enameled wire on the shank of a $5/32$-inch drill bit. Wind the turns as tightly as possible so that the choke will retain a coil shape after you slide it off the drill bit.

The tuning coil is wound of No. 22 wire (hookup wire with the plastic insulation removed) over a $7/32$-inch drill bit. It consists of four turns. The turns are spaced so that they occupy one quarter inch.

The small 4-turn coil is equivalent to L-2 in the CB tuner. The L-1 winding is a separate coil, made up of two turns of hookup wire formed as shown in the photo (page 125). This "coil" is soldered to I-1 and I-7 and is self-supporting. The coil is inductively coupled to L-2, and the coupling can be varied simply by bending the supporting leads.

### Checking the Unit

With the new coil in place, turn on the unit. You should hear the familiar "rushing" sound as was true with the CB unit. With an antenna connected—and assuming you are in a busy aircraft area—rotating the tuning capacitor slowly should pick up a signal. Often, these signals last only for a few seconds; aircraft communication is a marvel in concise conversation.

Because of the high frequency in use, the dimensions and shape of the coil are highly critical—even the length of the leads on the coil is important. For that reason, there is some chance that the circuit will tune some *other* band than the aircraft band. If you suspect this, first try squeezing the turns together until they almost touch; then rotate the capacitor. Should this not solve the problem, try "opening up" the coil, which decreases the inductance and allows tuning to a higher

frequency. With the coil one way or the other you should find the aircraft band.

## Testing With Another Receiver

Another approach can be used if you have access to a receiver which tunes the 108 MHz to 170 MHz band (which includes the 108–135 MHz aircraft band). You can use this receiver to "find" the band. Simply tune the vhf receiver to the middle of the aircraft segment on the dial and, with the tuner placed near the vhf receiver, turn the capacitor on the tuner. You should pick up a kind of garbled whistle (the superregenerative tuner will radiate a signal for a short distance).

If the tuner is generating the familiar rushing sound, but you cannot find it on the vhf set, chances are coil L-2 has too much inductance, and spreading the turns should solve the problem. If you find that the tuner covers only part of the aircraft band, increase the inductance (turns closer together) or decrease the inductance (turns farther apart) until you achieve the coverage you want.

## INTO THE CABINET

Once the unit is operating properly, installing it into the cabinet is the final step. The drawing shows the assembly.

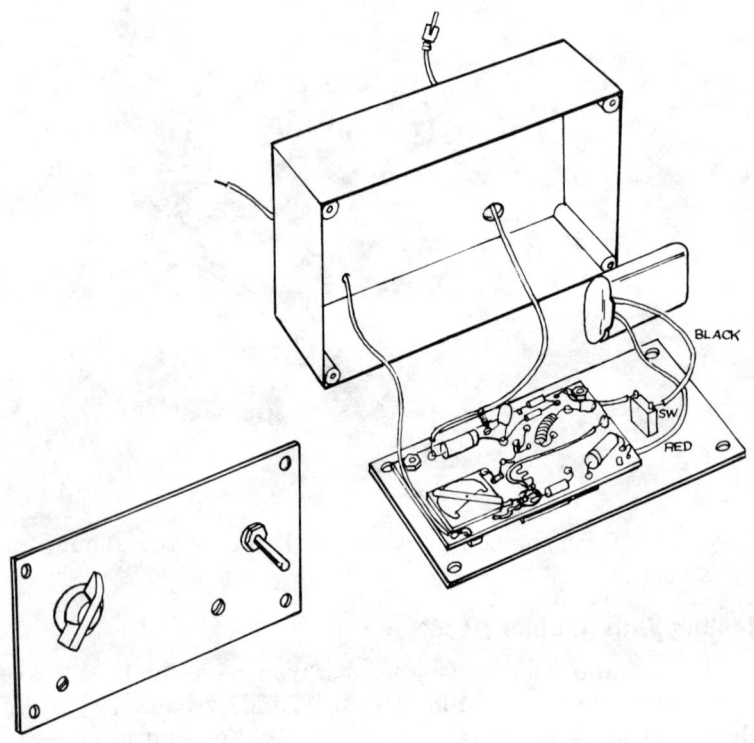

## ANTENNA AND ANTENNA COUPLING

A 2-foot (61-cm) length of wire (mounted vertically) may suffice for the aircraft band. For CB, an 8½-foot (2½-meter) piece, also as vertical as possible, is recommended. If you have an outside antenna, by all means try it.

For aircraft reception, try moving the small antenna coil (L-1) closer to or farther away from the tuning coil. The position for good reception may be quite critical.

**CHAPTER 12**

# Shortwave Receiver

At night in the standard am broadcast band, you may often pick up stations 1000 miles or more away, particularly during the winter months when static levels are low. But these distances are nothing compared with what you can do with a shortwave receiver, even a very simple one like that described in this chapter.

The unit shown in the photograph, tested in a basement workshop in Denver, pulled in powerful stations like Radio South Africa and Radio Moscow, with only a 10-foot (3-meter) piece of hookup wire for an antenna. With a good outdoor antenna, even weak stations rolled in from all over the world.

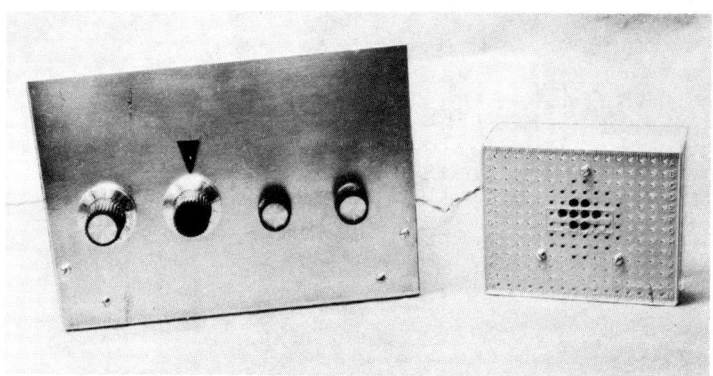

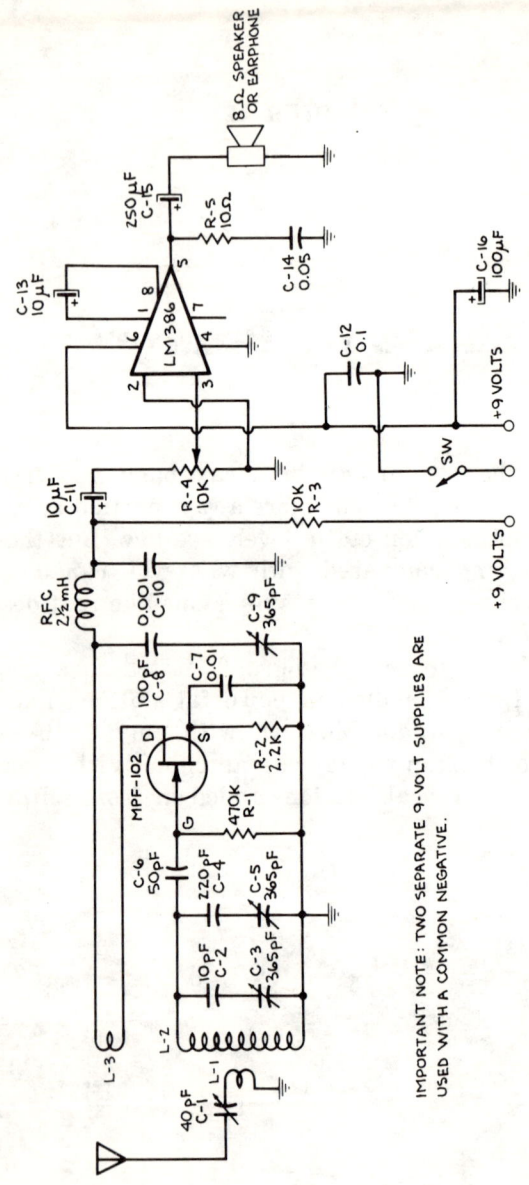

The circuit used is a bit cranky to operate—the penalty for achieving a lot of sensitivity with a minimum of parts—but once you get the hang of tuning it, the receiver can be the source of a lot of pleasure. To the writer, there is some kind of magic, and a glow of satisfaction, in listening to the Westminster Chimes of Big Ben in London tolling midnight, and hearing that magnificent sound on a small piece of equipment put together with one's own hands.

Like the Transistor AM Receiver, this set uses a transistor regenerative detector. The detector output is amplified by an IC audio stage. If this stage looks familiar, you are correct; it is the same amplifier circuit that we utilized in the chapter on the Loudspeaker Amplifier. In this unit, instead of joining two separate projects, we build them up on the same chassis.

In addition to enabling you to become an swl (shortwave listener), the set may be of interest to you for another reason. When operated with the detector deliberately set so that it oscillates, the receiver will enable you to listen to cw (code signals) within the amateur 7-megahertz band. These signals can be highly useful, particularly if you are practicing the radiotelegraph code in order to become a radio amateur.

Further, once you have become really expert at tuning the set, you can listen to the ssb (single sideband) amateur phone stations on 7 MHz. Tuning in the "sidebanders," as they are known in CB circles, takes patience and skill. Later in this chapter, you will learn how to go about it.

## COLLECTING THE PARTS

As always, the first step is to collect the parts you need. The shopping list covers only those parts for the regenerative-detector portion of the circuit. As mentioned, the amplifier is the IC circuit from a previous chapter. Some of the parts in the detector circuit are the same as those used in previously described units, which can be utilized if you want to "move up" to this more complicated set.

## WIRING THE DETECTOR BOARD

We will use the familiar punched circuit board. Insert the push-in terminals as follows:

## Shopping List

| Quantity | Description | Labeled in Diagrams |
|---|---|---|
| 1 | 40 pF trimmer capacitor | C-1 |
| 1 | 10 pF mica capacitor | C-2 |
| 2 | 365 pF air-dielectric variable capacitor (Calectro A-1-227 or equivalent) | C-3, C-5 |
| 1 | 220 pF mica capacitor | C-4 |
| 1 | 50 pF mica capacitor | C-6 |
| 1 | 0.01 µF Mylar capacitor | C-7 |
| 1 | 365 pF miniature solid-dielectric variable capacitor (Calectro A-1-233 or equivalent) | C-9 |
| 1 | 100 pF mica capacitor | C-8 |
| 1 | 0.001 Mylar capacitor | C-10 |
| 1 | 10 µF, 16 volt electrolytic capacitor | C-11 |
| 1 | 470K, ¼ watt resistor | R-1 |
| 1 | 2.2K, ¼ watt resistor | R-2 |
| 1 | 10K, ¼ watt resistor | R-3 |
| 1 | 10K volume control with switch | R-4 |
| 1 | 2.5 mH rf choke | RFC |
| 1 | Coil (see text) | L-1, L-2, L-3 |
| 1 | Transistor, MPF-102 or equivalent | |
| 1 | Perforated board, 2½ × 2¾ inches (6.4 × 7 cm) | |
| | Misc hardware | |

B-3, B-7
C-7
D-9
E-1, E-2, E-7, E-10, E-13
H-1
I-2
J-13
L-4, L-9, L-13
M-12

Two soldering lugs are bolted on the back side of the board at hole N-6.

If you have built some of the units described in previous chapters of this book, you should be able to tackle much of this project without step-by-step wiring instructions.

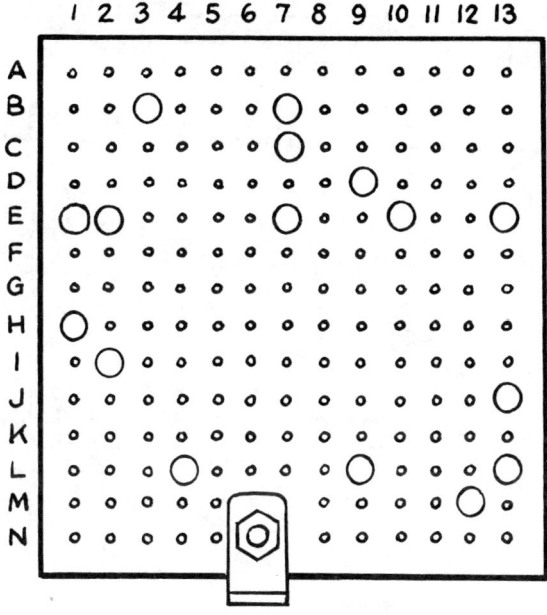

1. Check the board to see that the terminal clips and lugs are all in place.
2. Following the drawing, complete the wiring on the board. (Omit the coil connections for now.)
3. Turn the board over and solder the parts in place. (See drawing, page 132.) As always, use a heat sink to protect the transistor from heat during soldering.

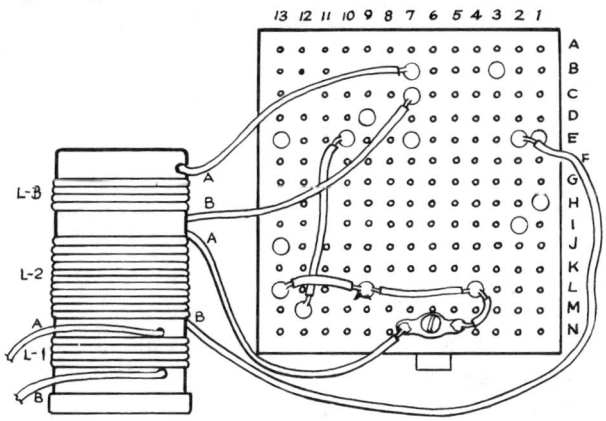

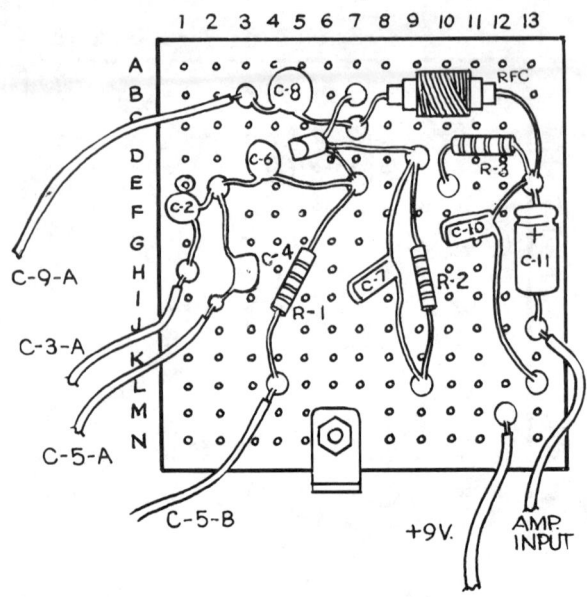

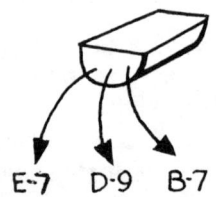

Check and recheck wiring. Okay? Lay the board aside while you make up the coil.

### WINDING THE COIL

The coil is wound on a plastic pill bottle, 1 inch (2.5 cm) in diameter and 2½ inches (6.4 cm) long. The technique is the same as that used in making the coil for the Two Hour radio, except that this time we use hookup wire (No. 22 solid). As before, drill holes through the bottle. Pull the wire through two holes to lock it in place, and then by rotating the coil form, wind on the specified turns for each coil.

| L-1 | Antenna coil | 4 turns |
| L-2 | Tuning coil | 10 turns |
| L-3 | Tickler coil | 3 turns |

Be certain to leave leads long enough so that you can make the needed connections to the circuit board.

### PREPARING THE CHASSIS AND PANEL

The metal chassis is 2 × 5 × 9 inches (5.1 × 12.7 × 22.9 cm). The panel is a piece of copperclad board of the type used for printed circuits; it measures 6 × 9 inches (15.2 × 22.9 cm).

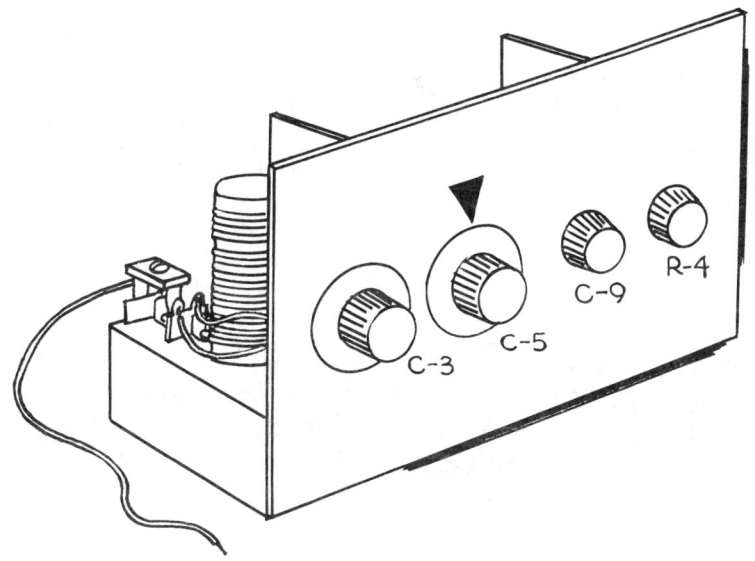

Drilling the panel will be governed by the components used. Exact placement of the variable capacitors is not important, but spacing the components equidistant from each other will make the unit more attractive. The "pointer" is a small piece of electrician's tape.

The two air-dielectric variable capacitors are bolted to the chassis, as is the front panel. Similarly, the solid-dielectric variable capacitor and the volume control go on the panel. ("Dielectric" refers to what is between the moving and fixed

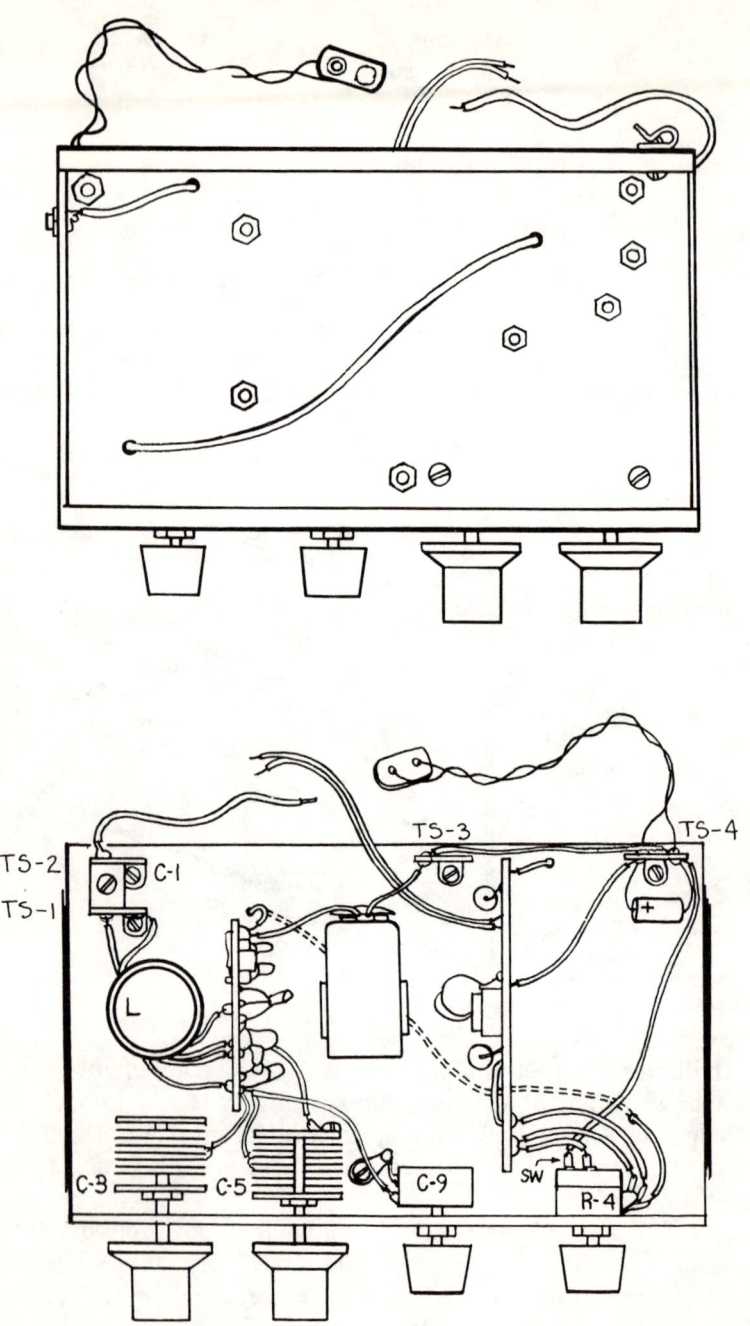

plates of the capacitor, air in one case and a thin plastic in the other.)

Four terminal-lug strips are bolted to the rear of the chassis as shown in the illustrations. Strips TS-1 (one lug insulated, one grounded) and TS-2 (one insulated lug) are for mounting C-1. Strips TS-3 (one insulated lug) and TS-4 (two insulated lugs) are for battery connections.

Many times in electronics, the connection to a chassis—or to a large "ground" area on a printed circuit board—is actually part of the wiring, and if such a connection is omitted, the unit will not operate. Note that in this set the amplifier board is grounded to the chassis via a mounting angle, as is the detector board. Capacitors C-3 and C-5, having metal frames, are grounded when they are bolted to the chassis.

### ASSEMBLY

In a unit of this type, there is an advantage to checking out the stages separately. With that in mind, we will be certain the amplifier stage is working before we wire in the detector stage.

First, build up the amplifier on a board as outlined in Chapter 5 and the diagram on page 136. Compare this diagram with the unit in Chapter 5, and notice that the volume control is omitted, as is one capacitor. These parts are wired in during final assembly. The board is mounted on small angles.

Note that three soldering lugs are held by the bolt through hole N-14, which secures the board to an angle. There are two lugs on the "wiring" side of the chassis and one on the opposite side. As mentioned, bolting the angle to the chassis establishes the "ground" connection. The angle on the other end of the board helps support it, but it has no function in the wiring, being bolted to hole N-1 in the panel, which has no circuit parts associated with it.

The next drawing (page 137) shows the assembly wiring after the board is in place.

- ☐ Connect the battery connector and capacitor C-16 (observing polarity). Wire in the lead which connects SW-2 to TS-4.
- ☐ Mount a holder for the 9-volt battery on the chassis.

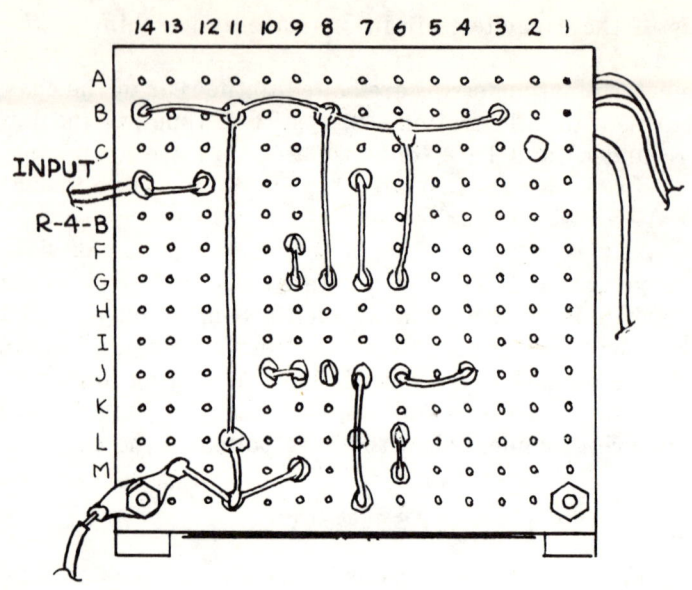
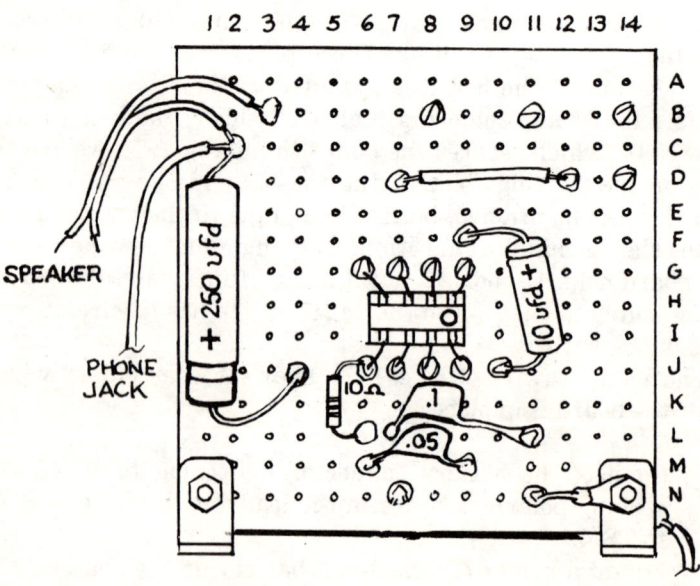

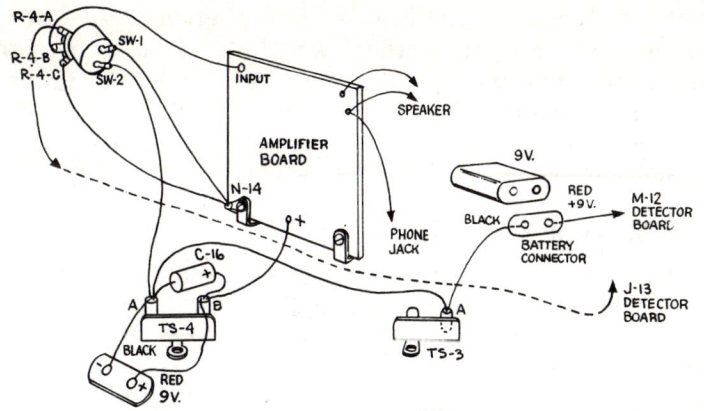

☐ Wire SW-1 to the grounded lug on the amplifier board, N-14, and R-4-C to the same point.
☐ Hook the speaker leads to points C-2 and B-3. If a phone jack is to be provided, run a lead from point C-2 to the "hot" terminal on the phone jack. The jack *must* be the open-circuit type—a closed-circuit jack will *short the output to the chassis when the plug is removed!*
☐ Connect the black lead of a 9-volt battery connector to TS-3-A.
☐ Wire lug B on TS-4 to N-7 on the amplifier board.

Yes, the set uses *two* battery sources. Ideally, the amplifier stage should be supplied from six 1½-volt "AA" cells, which add up to 9 volts. This combination will give *much* longer life than a small 9-volt transistor battery. The latter is fine, though, for the detector stage, which draws little current.

### TESTING THE AMPLIFIER

Connect the speaker to the speaker leads. With the switch off, hook up the 9-volt amplifier battery. We will check out the amplifier stage first, using a technique familiar to any service technician.

**Signal Injection**

One basic servicing technique in all electronics is that of introducing a signal at the input of a circuit and seeing if it comes out (usually amplified) at the other end. To test our amplifier, we can utilize the Two Hour radio as a signal source; with two wires, apply one lead from it to terminal D-14 and the other to the chassis. Snap on the switch. If all is well, you will hear the amplified signal as described in Chapter 5 when we hooked the Two Hour radio to the amplifier.

### Rough 'n Ready Signal Source

Should the Two Hour radio not be available, there is another rough test which professional service technicians often use, even if they have a bench full of test equipment. Simply applying a forefinger to a screwdriver and touching the screwdriver to D-14 should yield a strong hum. What is happening is that a person's body will pick up induced ac from the power lines in the house, and this is amplified, just as a signal from a tuner would be amplified.

### What You Know Now

By now you have gained the obvious advantage of testing the amplifier stage *before* completing the set. Later, should the set be nonfunctioning, you will know that the trouble is in the detector board, and you can direct your attention to it. This unit-by-unit approach is highly useful in building more complex equipment, just as it is in testing a tv set, stereo, etc., stage by stage.

### Signal Injection and Signal Tracing

There are two basic techniques used for most servicing of electronic equipment: "signal injection" (the method outlined above) and "signal tracing," which requires a bit more equipment. The idea with signal tracing is to "tap" the unit under test at specific points and find out where in the sequence of stages the signal is lost. This is usually done with an audio amplifier (like the utility amplifier described earlier) or with an oscilloscope, which displays the signal on a cathode-ray tube.

## COMPLETING THE SET

If your amplifier is working, you are ready to complete the assembly by adding the detector board.

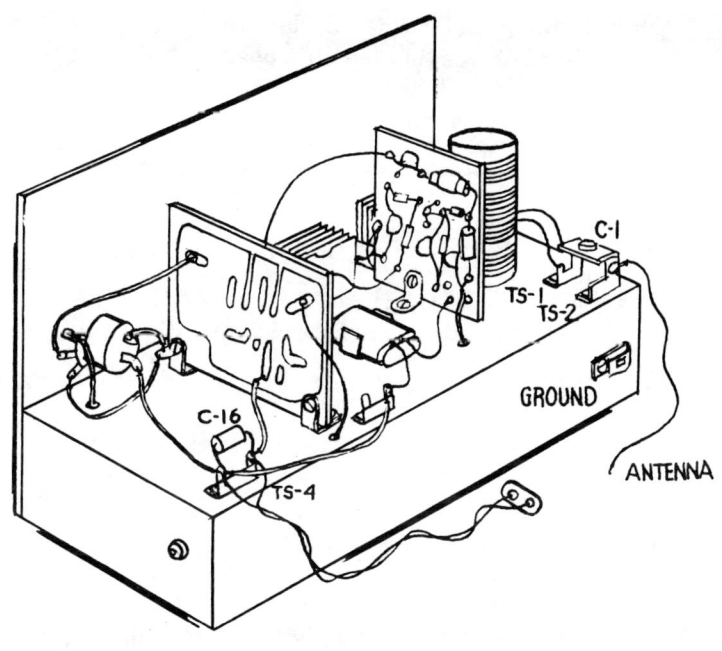

- ☐ Mount the detector circuit board to the chassis by means of a small angle. Bolt it down securely, because this establishes a ground connection to the chassis.
- ☐ With a single bolt, secure the pill-bottle lid (for the coil wound on a pill bottle) to the chassis.
- ☐ Bolt a Fahnestock clip (which provides for a ground connection) to the chassis.

This completes the parts mounting. Next—

- ☐ C-9 is the capacitor closest to the volume control. From the lug connection to the stationary plates, run a lead to B-3 on the detector board.
- ☐ Connect the other lug on C-9 to a soldering lug bolted to the chassis.
- ☐ From a soldering lug on the frame of C-5 (C-5-B), connect a lead to L-4 on the detector board.
- ☐ Connect the "stator" lug on the same capacitor (C-5-A) to I-2 on the detector board.
- ☐ From the stator (stationary plates) lug on C-3 (C-3-A) run a lead to H-1 on the detector board.

☐ Connect the red lead from the transistor-battery connector to M-12 on the detector board.
☐ From J-13 on the detector board, run a wire lead through a hole underneath the chassis to another hole near the volume control. The wire (insulated hookup wire, of course) surfaces at this point and is connected to R-4-A (see illustrations).
☐ Connect R-4-B to D-14 on the amplifier board.

We are almost finished with the wiring. Adding the coil and installing the antenna trimmer capacitor are the final steps.

☐ Mount the pill-bottle coil form by snapping it into the lid bolted to the chassis.
☐ Following the drawing shown earlier in the chapter, connect the coil to the board. For purposes of illustration, the coil is shown set off to one side. Actually, the leads between the coil and the board should be as short as possible.
☐ Wire L-1-A and L-1-B to the lugs on TS-1. Lead L-1-A goes to the lug which grounds to the chassis. L-1-B goes to the insulated lug.
☐ Trimmer capacitor C-1 is soldered between the insulated lug on TS-1 and the insulated lug on TS-2. A short length of wire is soldered to TS-2 to provide connection to an antenna.

### FINAL CHECKOUT

Look over all the wiring and soldered connections. Check them against the illustrations as well as the text. If possible, have someone else do the checking as well—for some reason, it is always easier to find *other* people's mistakes. *If* you are satisfied, plug the IC into the socket on the amplifier board. Be certain to place it so that the dot on top of the IC is *toward* the front panel.

Connect an 8-ohm speaker (or earphone) to the speaker leads (or phone jack). Snap the battery connectors onto the batteries. Rotate the volume control clockwise. You should have some sound from the speaker. Now, using the "finger on the screwdriver" signal injection method described earlier, touch J-13 on the detector board. You should hear a hum.

Rotate the volume control. The hum should decrease as you turn it counterclockwise.

So far, so good. Hook whatever antenna you have to the antenna wire on TS-2.

With a screwdriver, adjust the trimmer capacitor about halfway open. Now rotate C-9 (regeneration control) until you hear a soft "plop," perhaps accompanied by a gentle humming sound that indicates the detector is oscillating.

Rotate the main tuning capacitor (C-5) until you hear a strong whistle. This indicates that you are in tune with a station. Now—carefully—back off the regeneration control. If you are in tune with an am (amplitude modulated) shortwave station, at a certain point the whistle will disappear and the station will become intelligible.

You will discover that tuning with C-5 is extremely critical. For that reason, set C-3 about halfway open, and then find the station again. Now, tune the station in carefully with C-3, which tunes quite broadly. In technical terms, C-5 is the *band-setting* capacitor, and C-3 provides *bandspread*.

## LEARN BY PRACTICE

The foregoing sounds more complicated than it actually is. Half an hour of experimenting is better than hundreds of words, but perhaps a quick summary will help.

1. Rotate the regeneration control until the set begins oscillating. This may require either clockwise or counterclockwise turning, depending on the capacitor used.
2. Do the approximate tuning with C-5.
3. Tune the station in with C-3.

You will find that the set is most sensitive for receiving broadcast stations just before the point where oscillation begins. Selectivity (ability to separate stations) is best at this point, also. As you find the channels of powerful broadcasting stations, selectivity becomes an important factor.

## SOME FINE POINTS

Trimmer capacitor C-1 allows adjusting the antenna coupling to different antennas. The tighter the coupling (screw turned in), the stronger the received signals will be. However, if the process is carried too far, selectivity suffers, and "dead spots," where it is impossible for regeneration to occur, may appear in the tuning range. The cure is to reduce the coupling.

## TUNING FOR CW

Whereas with am signals the set should be operated just below the point of oscillation, for reception of cw (continuous-wave code) the set is tuned with the regeneration control advanced just beyond the point where oscillation begins. High-speed code signals are scattered throughout the tuning range of the set. However, if you want to listen to cw for code practice, the 40-meter amateur cw band (7000–7150 kHz) is best.

The trick will be to find the band. As a starting point—assuming you have variable capacitors like those illustrated—adjust C-5 so that the moveable plates extend only about $3/8$ inch (measured at the tip end) above the stationary plates. Now, with the regeneration control advanced so that the set is in an oscillating condition, tune slowly up and down the band. If you started fairly close to the correct frequency, you should hear code signals with dots and dashes in the form of a nice, clean, high-pitched whistle, the tone of which you can vary by fine tuning with C-3. The stations sending more slowly will be those in the "Novice" portion of the band.

## SSB, ALSO

If you have successfully tuned in cw stations, you are ready to tune in single sideband (ssb) amateur phone stations. This requires patience and a steady hand, but you can learn to do it.

Assuming that you have found the amateur 40-meter (7 MHz) cw band as described earlier, turn C-5 slowly to a more open position. You should begin to encounter some voices speaking a kind of incomprehensible gibberish. Pick a signal of moderate strength and then very, very slowly turn C-3. As you tune across the signal, suddenly the voice will become understandable. The point at which the signal turns from a whistle or gibberish into something you can understand is critical, but with a little practice you can find it. Tuning is easiest in the daytime, when fewer stations are on the air.

If stations are extremely strong, you may have trouble tuning them in. The remedy is to reduce the antenna coupling by opening up trimmer capacitor C-1.

## TROUBLESHOOTING

If the set does not operate as described, the initial step is to check and recheck all the wiring and soldered connections. We have already determined that the amplifier stage is operating, so we know that the trouble has to be in the detector board or associated wiring.

The first thing to determine is whether the detector stage will oscillate when the regeneration control is advanced.

1. Back off the screw so that C-1 is wide open.
2. Adjust C-5 so that it is about 80 percent open.
3. Turn the volume control all the way clockwise.
4. Turn the regeneration control fully counterclockwise.
5. With your hand, grasp the coil. You should hear a soft "plop" as you touch the coil. If you do *not* hear it, try turning the regeneration control all the way clockwise. (Unfortunately, which way this control operates will depend on the brand of capacitor in use.)
6. No "plop"? Then, assuming there is no wiring error or bad soldered connection, it may be that the transistor in use requires a bit more "feedback." To remedy this, increase "tickler" winding L-3 to 4 turns.
7. Now a "plop"? Turn capacitor C-5 step by step toward the fully closed position, checking as you move down in frequency with the "grasp-the-coil" technique. If oscillation is lost partway down, increase L-3 to 5 turns.
8. Okay? Increase the antenna coupling by tightening the screw in C-1. This gives louder signals. Too much antenna coupling may result in "dead spots" on the C-5 dial. The remedy is to back off the coupling.

Once you have the set oscillating, you have the sensitivity which regeneration provides—*far* more than with any other simple circuit within the sw range. As mentioned earlier, the trade-off for achieving this simplicity is that operating the set takes some tuning skill, arrived at by practice.

**CHAPTER 13**

# Making Your Own Printed Circuits

The "wiring" in much of the electronic equipment in use today utilizes a technique known as "printed circuits." Thin boards made of material similar to the perforated boards used for much of the construction in this book are coated with a layer of copper. Then, by a photographic process, the "wiring" is printed on the board. Finally, excess copper is removed with an etching solution, and what remains provides the various leads required to connect components together. The photo shows a "conventional" hand-wired circuit board with an equivalent homemade printed circuit board just below it. Note how the same layout was followed.

The advantages of printed circuits are many. For example, once a "negative" is available, many identical boards can be mass produced. If the original board has the correct wiring pattern, so do all of the duplicates, thus reducing the possibility of wiring errors. Further, printed circuit boards allow compact layouts, which are less likely to cause feedback and other problems.

Very important to the electronics fan who likes to build equipment, it is now common practice for technical magazines to print patterns for printed circuit boards, thus enabling the builder to duplicate exactly a unit described in a technical article. Often, the actual printed circuit boards are

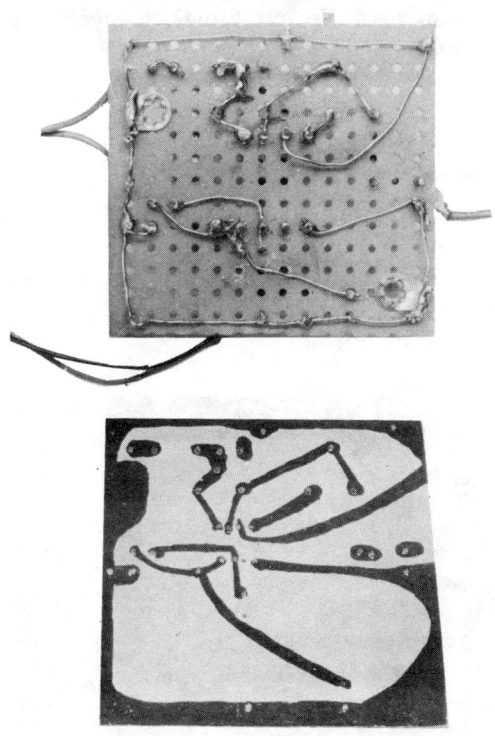

available from mail-order or other sources so that the builder can construct the unit in a minimum of time simply by buying the boards and soldering the components in place.

## MAKING YOUR OWN

Because of the advantages of printed circuits, half a dozen different methods of making them are available to the builder. Kits of material are on sale by many electronic parts companies, and it is common practice for the kits to include instructions for utilizing the particular method for which the kit was designed.

## THE SIMPLEST APPROACH

A method which the writer uses for creating simple printed circuit boards is *not* sold in kit form, perhaps because so few items are necessary: a bottle or can of etching material, a special kind of pen (readily available), a piece of copperclad board, and a scouring pad of the type used to scrub pots and pans in the kitchen.

The best way to learn the technique is to use it. As a project, we will select the amplifier board from the shortwave receiver, which was intentionally laid out in such a way that it can be translated into a printed circuit (pc) board quite easily.

To start, we need to make a copy of the pc pattern. This can be done in most offices or libraries, or even in a supermarket or drug store that has a coin-operated copying machine. The full-size pattern is shown here.

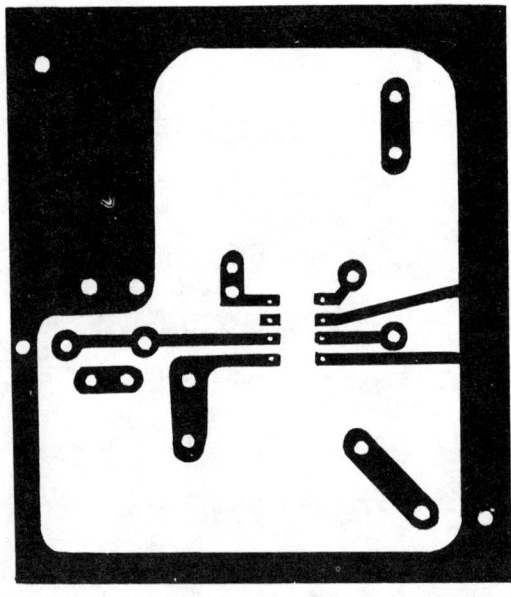

Now, for a brief description of the method. Printed circuit boards are produced by protecting part of the area of the copper from the etching solution, by any one of several methods. The most direct way is to "draw" the printed circuit layout directly on the copper-plated board with etchant-resistant ink (actually a kind of lacquer). This can be done with a brush, but unless you have some art skill, the result can be rather bad. The alternative is to use a pen specially made for the purpose. Somehow—perhaps because we all are more practiced in handwriting than in painting with a brush—the pen approach is far better. Even so, it is difficult to "draw" without some kind of guide points, and that is where the method in this chapter differs from the techniques commonly used. In most methods, the printed circuit board is created, and *then* the holes are drilled for the parts. For our first board, we will drill the holes *first* and use them as guide points for drawing. The method is much the same as that used in children's books in which the child runs a line from one number to another, and in the process creates a drawing.

We'll assume that we have a full-size copy of the pattern. With a hacksaw or other fine-tooth metal saw, cut a board of the correct dimensions to match the pattern. Then tape the pattern to the copper side of the board.

With the pattern firmly in place, drill the holes. A number 60 bit, 0.0400 inch in diameter, is a good one to use. Calectro (GC Electronics) puts out a small package of three bits (catalog number 22-275) which includes this size.

If you have a small drill press available to you, the drilling is easy. The job is a bit more difficult with a hand drill, because the bit is tiny and there is danger that you will tip the hand drill and bend the bit. However, if you are careful, the job can be done.

Locate the "holes" in the pattern. Punch through the paper pattern with the end of a compass, awl, or other tool that has a very sharp point. This is to make it easier to start the drill accurately. Be especially careful when locating the eight holes for mounting the socket. These must be "on the nose"—the other holes are less critical as to location. Practice on some of the others first. With the board placed on a block of wood, and with the edges taped to it, start drilling. If you are using a hand drill, "rock" the large gear back and forth to start the hole. Then, being careful to hold the drill vertically, drill *through* the copper surface. You can drill all the way through the board if you wish, or do that job later, whichever seems easier.

In the case of the large hole in one corner of the pc board, drill in the center of the desired hole area. This hole can be enlarged later so that it can handle a 6-32 bolt.

As the final step, drill the eight socket holes. Again, be very careful. Unless these holes are in the right place, the socket will not slip into them.

Be certain you have drilled *all* the holes. It is embarrassing to pull off the pattern and discover you have missed a couple of them.

## APPLYING THE PATTERN TO THE BOARD

From here on we need to treat our board with respect. If it is dirty, it will not accept the resist lacquer or etch properly. So, the next step is to clean it with a soap-filled scouring pad. Keep at it until the copper is shiny. Then rinse the board thoroughly to remove the soap. Handle the board by the edges—you don't want to destroy the cleaning job with finger prints. Dry the surface of the board with lint-free paper toweling, or simply wait until it *is* dry.

Now we will use our etch-resist–ink pen (Calectro J4-629 is one type). Shake it a few times, and try it out on a piece of paper.

Place the board on a piece of paper toweling on a bench or table.

The idea is to draw in the elements of the pattern, using the holes as guide points. Start with the simple elements, for example, one of the three in which there is an oval-shaped area with a hole on either end.

In using the pen, a *very* light touch is desirable; in fact, the weight of the pen alone is enough. The idea is to *flow* the ink onto the surface, not to write on it as if you were writing with a pen. The writer has found it works best to use a circular motion, gradually building up the coated area—and taking time to do it *slowly*. There is no danger of putting too much ink on the surface, but if you don't put on enough, the resulting board may have pit marks. Unless these are so large that they break the continuity of the wiring, there will be no problem, but for the sake of good craftsmanship, do the best job you can.

Do *not* go back and try to re-cover a spot already covered while the surface is still wet. If you do this, you will "lift" the resistant ink and make matters worse. Spots you miss can be touched up later after the resist is dry.

As before, the socket is the critical area. Be certain to apply enough etchant so that a good lead will result, but avoid overdoing it, or you may run one socket lead into contact with another. Should this happen, wait until the resist dries, and then scrape it off with a small knife or a razor blade.

Check the drawing on the board against the pattern in the book to be certain that it is all there. Exact shapes are not important. The important thing is that all of the holes in the board are connected together as shown.

## ETCHING THE BOARD

For the etching job, we need some etching solution (Calectro J4-628 is typical); a small glass, plastic, rubber, or aluminum tray; and a small piece of wood such as a $\frac{1}{4}$-inch dowel sharpened on one end or an "orange stick" of the type used for manicuring.

As is usually true with chemical actions, the etching process is affected by temperature. For that reason, warm the solution by placing the container in a pan of warm water out of the faucet. In addition, warm up the etching pan or dish in the same fashion.

Be *careful* not to splash the etching solution onto skin or clothes. As the label on the can will show, the solution must be treated with respect. The etching solution will stain *anything* it touches, so it is a very good idea to wear rubber or inexpensive plastic gloves while working with it.

Place the board in the etching tray pattern side up. Pour in enough etching solution to cover the board by about $\frac{1}{4}$ inch. *Gently* rock the etching tray back and forth every couple of minutes, and from time to time lift the board up with the wooden stick. The idea is to distribute the etching solution over the board, and to assist in carrying away the excess material.

The etching process actually starts immediately, but it will be some time before you can see any evidence of it. The writer's observation is that the directions which come with etching solutions are overly optimistic—if the etching time is supposedly 20 minutes, it may turn out to be more like 40 minutes. There are a lot of variables, of course: thickness of copper coating, temperature, strength and age of solution, etc. After about 15 minutes with a typical solution, you will begin to see the copper pull away from the larger areas, and the board begins to appear.

## HOW TO KNOW WHEN YOU ARE DONE

Once you see that the etching process is proceeding in earnest, check the board every 5 minutes by lifting it out of the solution and washing it off with running water. When it appears that there is nothing left in the "open" areas but the plastic board, take one more look, tilting the board back and forth under a light to see if any copper remains. If you can spot none, you are finished so far as the etching process is concerned.

## FINAL CLEANUP

Now we will use the scouring pad again, rubbing the pad on the circuit-board pattern until the black ink is all removed and the remaining lines of copper are shiny bright. Wash the board to remove any trace of soap.

Examine the holes in the board. If you did not drill them all the way through the first time, that is the next thing you should do.

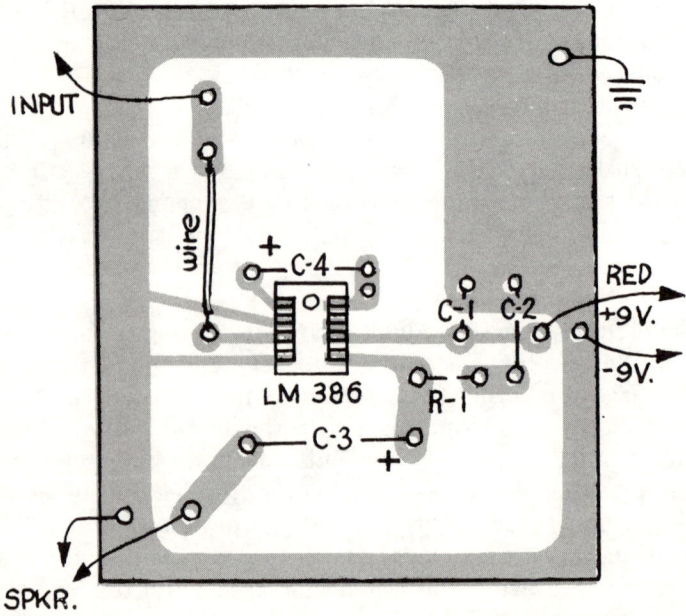

## WIRING A CIRCUIT BOARD

The common practice for illustrating circuit boards in technical magazines is to show a kind of X-ray view of the board, with electronic symbols or nomenclature such as "C-1" indicated to show placement of parts. The illustration on page 150 shows the parts which are to be added to our board.

To add parts, push the leads through the board, and bend them over. (For easier soldering, scraping the leads with a knife before doing this is a *good* idea.) Then—quickly and

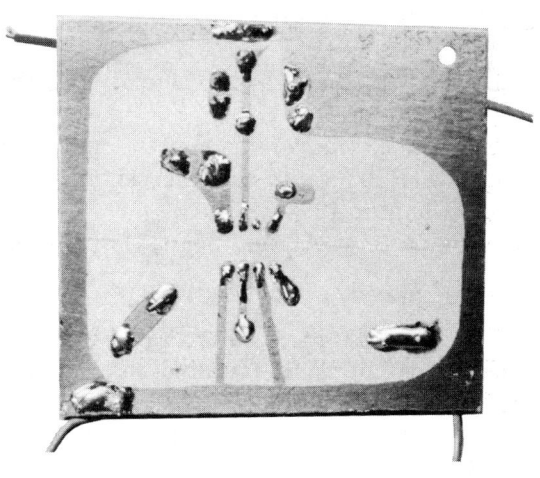

carefully—solder each lead to the proper point on the circuit board. The photos show the board from top and bottom. Clip off excess wire. Be certain no solder has run between circuit leads; if it has, remove it with a small, sharp knife.

Use the *minimum* amount of solder necessary to make a good connection, and avoid holding the iron to the board so long that there is danger of damaging the small copper leads. A light iron (about 27 watts) will make this easier to do, although an iron of somewhat higher wattage can be used if you are cautious. Do *not* use a soldering gun; doing so is almost certain to damage the printed circuit.

The parts to be mounted on the board are as follows:

| *Labeled in Diagram* | *Description* |
|---|---|
| C-1 | 0.1 $\mu$F, 16 volt Mylar or ceramic capacitor |
| C-2 | 0.05 $\mu$F, 16 volt Mylar or ceramic capacitor |
| C-3 | 250 $\mu$F, 15 volt electrolytic capacitor |
| C-4 | 10 $\mu$F, 15 volt electrolytic capacitor |
| R-1 | 10 ohm, $\frac{1}{4}$ watt resistor |
| LM386 | Integrated circuit, Type LM386 or equivalent |

These parts are identical to those used on the amplifier board of the Shortwave Receiver (and in the Loudspeaker Amplifier).

The test procedure is also the same as the one we used before:

1. Insert the IC into the socket. The end with the dot goes toward capacitor C-4.
2. Hook the stage to the volume control and the speaker.
3. Use the Two Hour radio or other signal source as described in Chapter 12.
4. Connect the battery. At this point, the unit should come alive.

If all is well, the printed circuit board can be used in the shortwave set instead of the perforated-board version. For all practical purposes, the results will be the same, and if you have not yet built the shortwave receiver, you may want to use the printed circuit board the first time around.

The photo shows the board in place in the receiver. As with the perforated-board amplifier, the "ground" to the chassis is provided by the mounting brackets.

# CHAPTER 14

# Parts and Parts Substitutes

Because most transistors and ICs operate with little heat buildup, modern electronic equipment requires far less servicing than was true in the days of radio tubes. Because less servicing is needed, fewer parts are available from local suppliers—sometimes including parts which are useful in homemade equipment.

For this book, a special effort was made to design units that use standard components usually available from many sources. However, to make it easier for you to obtain the part you need in your city, this chapter is devoted to discussing parts substitution and some suggestions for ordering by mail if that is a better answer for you.

## OVERALL CONSIDERATIONS

First, we will look at some overall suggestions.

### Fixed Capacitors

There are three general types of fixed capacitors:

1. Larger capacitance electrolytic units (10 $\mu$F to 300 $\mu$F) often used for filtering, although sometimes used for signal-carrying circuitry.
2. Medium-range bypass capacitors (for example, 0.01 $\mu$F to 0.2 $\mu$F).
3. Rf capacitors in the range 5 pF to 1000 pF.

*Electrolytics*—If a size of 20 μF or lower is specified, try to obtain the correct size, which should be readily available. In the range 50 μF to 300 μF, going to a component of higher capacitance is usually perfectly acceptable; for example, if the parts list specifies 200 μF, a 250-μF unit will work fine.

*Bypass*—Again, going to the next larger size ordinarily is okay. For example, a 0.02-μF unit can be substituted for a 0.01-μF unit. Bypass capacitors usually are either Mylar or ceramic. For the circuits in this book, whichever type is available from your supplier can be used.

*RF Capacitors*—Capacitors that carry the radio-frequency portion of a circuit are more critical and should be the correct size. They should be either silver mica or polystyrene. You may find the same value in a ceramic capacitor, but use this type only as a last resort.

**Variable Capacitors**

The units in this book specify 365 pF variable capacitors. These come in various types. Except for the CB/Aircraft tuner, they are all interchangeable. However, there are some special considerations that will be covered later in this chapter.

**Resistors**

Resistors, as you have already learned, are rated according to both resistance and power capacity. As an example, a resistor may be specified as 100,000 ohms, ¼ watt. A resistor with a higher wattage rating can be used instead. For example, a 100,000-ohm (100K), ½-watt resistor can be substituted for a 100K, ¼-watt unit. Unless the circuit requires a higher wattage rating, the smaller resistor is usually preferable for space reasons.

In the kind of circuits covered in this book, exact resistor values are not critical; anything within 10% may be used. For example, a 470,000-ohm resistor may be substituted for a 500,000-ohm resistor.

### Transistors and ICs

It is necessary to use the *exact* component in the case of an IC. For transistors, it is best to use the type specified, although substitution is usually possible. The MPF-102 specified for several projects is widely available. However, a 2N4416 should be a satisfactory substitute. For other substitutions, the counter man at any parts store will generally have a cross-reference manual and can help you buy the proper substitute.

## CONSIDERATIONS BY PROJECT

The foregoing paragraphs provide a general guide for parts substitution. Now we will consider the projects individually.

### Two Hour Radio

The Two Hour radio has few parts, so only a few substitutions are possible.

The 1N34 diode is a standard component, and the interchange manuals will indicate a number of substitutions. Actually, almost *any* diode will work in this simple circuit.

The tuning capacitor can be an air-dielectric variable of 365 pF capacitance. Simply be certain that the rotor of the capacitor goes to the same point in the wiring as C-1-A and that the stationary plates connect to point C-1-B.

The capacitor can be mounted on top of the circuit board, and a cardboard dial can be cemented to the capacitor. The accompanying photo illustrates a Two Hour radio modified in this way.

There is another possibility for the variable capacitor: using a two-gang miniature tuning capacitor like that shown in the next photo. Ordinarily, these capacitors have two sections.

We can parallel these and achieve enough capacitance to make them usable. Note that the two outer terminals are connected together to provide the "stator" connection. Capacitors of this type can be salvaged from "junked" small transistor radios, which will usually yield a diode as well, plus such items as earphone jacks.

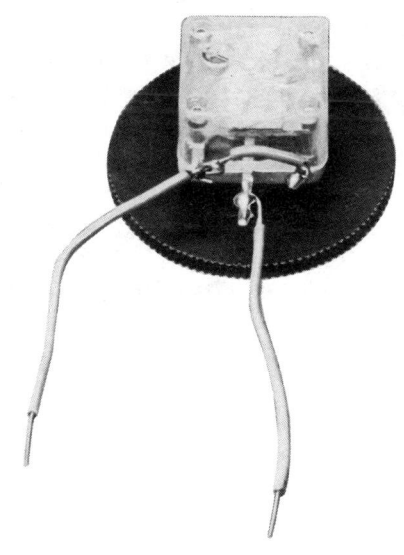

### Selective AM Tuner

The comments about capacitors and diodes made in connection with the Two Hour radio apply equally well to the Selective Tuner.

Tuning coils L-1 and L-2 with adjustable slugs are highly desirable. However, flat ferrite-bar coils can be used instead. The disadvantage of the nonadjustable coils is that there is no easy way to ensure that both of the dials read the same, because one coil must have *less* inductance than the other. This *can* be done by removing turns from the antenna coil a few turns at a time until the dials are in sync. Be careful in modifying the coil; the Litz wire ordinarily used is very tiny and difficult to solder without burning it up.

If you use the flat coils (such as Calectro D-1848), lay them side by side to ensure good inductive coupling. The coils can be secured with string run through the holes in the board. Do *not* use wire, which could introduce losses.

### Loudspeaker Amplifier

All parts for the Loudspeaker Amplifier should be widely available.

### Transistor AM Receiver

Most of the parts for the Transistor AM Receiver—with the possible exception of the coil and variable capacitors—are so standard that they should be available at any parts house. The comments regarding coils and variable capacitors made earlier in the chapter apply: utilize any 365 pF variable capacitor available, and a flat-bar coil can be used as well. Wind the tickler coil directly over the coil. Some experimenting may be necessary. Use as *few* turns on the tickler as will allow the set to break into oscillation with the regeneration control advanced.

As shown in the circuit diagram, the regular antenna connection is made to the tap on the coil, whether it be the recommended adjustable-core coil or a nonadjustable coil. This tap is provided for use with some types of transistors, particularly the older variety.

For this tuner—as with the previous two—binding posts or terminal strips can be substituted for the Fahnestock clips.

Having trouble obtaining a transformer? A 10K resistor can be substituted for the transformer winding, but this will result in less output and may require more turns on the tickler coil to ensure regeneration.

### Hi-Fi AM Tuner

Again, the parts for the Hi-Fi AM Tuner should be easy to find, with the possible exception of the coil, capacitor, and IC. Coil substitution was described earlier. Likewise, an air-dielectric variable capacitor will work fine, but it requires some modification of the board and case to allow for the larger size. For the ZN414, see the information at the end of the chapter regarding mail-order sources.

### Aircraft-CB Tuner

For the aircraft and CB tuners, layout is very important. Try to build the unit exactly as shown, which includes using the variable capacitor illustrated. To obtain small capacitance, the tuning capacitor in this circuit consists of a small fixed capacitor in series with a standard 365-pF variable unit. A midget 20-pF variable capacitor can be substituted, but it will require some modification of the layout. If you use the small variable capacitor, connect the coil from the stator to the rotor with the *shortest* possible connection. Some modification of the coil (by squeezing the turns together or pulling them apart) may be necessary to tune the desired band.

### Utility Amplifier

You should have no problem finding components for the Utility Amplifier at any well-stocked radio parts store. This includes the IC, which is made by two different manufacturers.

### Shortwave Receiver

The comments regarding variable capacitors given previously in this chapter apply also with respect to the Shortwave Receiver. All the other components, including the MPF-102, are widely available.

### Printed Circuits

Printed circuit supplies are available almost everywhere, including the many Radio Shack outlets.

## WHAT TO DO IF YOU CANNOT FIND PARTS LOCALLY

It is always best to try to buy parts from your local supplier, who can give you advice and assistance and often can answer questions regarding matters which you do not quite understand. However, if you live in a small town, there may be no electronic parts store nearby. Should this be the case, you can order parts by mail.

A large midwestern company which does business throughout the USA is

>Burstein Applebee
>3199 Mercier Street
>Kansas City, MO 64111

B-A stocks many parts and issues a free catalog, which you can obtain by writing for it.

A small company which will endeavor to stock the less common parts needed for building the units described in this book is the

>Johnson Company
>855 South Fillmore Street
>Denver, CO 80209

If you have difficulty obtaining variable capacitors, crystal earphones, transistors, ICs, or coils, send a stamp to the Johnson Company for their catalog. The firm is also planning to supply printed circuit boards for several of the projects.

As we go to press, the writer has just received a catalog from Radiokit. It contains a number of parts of interest to the readers of this book, plus many hard-to-find items needed by the radio amateur who likes to build equipment. For your copy of the catalog, send 25 cents to:

>Radiokit
>Box 411Q
>Greenville, NH 03048